UK price
£8.99

COMBAT AIRCRAFT
AH-64

Published by Salamander Books Limited
LONDON • NEW YORK

A Salamander Book

Published by Salamander Books Ltd.,
129-137 York Way,
London N7 9LG,
United Kingdom.

This edition © Salamander Books Ltd., 1992

ISBN 0 86101 675 0

Distributed in the United Kingdom by
Hodder & Stoughton Services
PO Box 6, Mill Road
Dunton Green, Sevenoaks,
Kent TN13 2XX.

All rights reserved. No part of this book may be reproduced, stored in a retrieval system or transmitted in any form or by any means, electronic, mechanical, photocopying, recording or otherwise, without the prior permission of Salamander Books Ltd.

All correspondence concerning the content of this volume should be addressed to Salamander Books Ltd.

Contents

Development	4
Structure	16
Powerplant	22
Avionics	26
Armament	32
Deployment	42
Performance and Handling	54
Improving the Breed	60
Specifications and Data	64

Acknowledgements

The author and publishers are grateful to all those who have helped in the preparation of this book. Sources of individual photographs are credited at the end of the book, but particular thanks are due to Robert Mack and Hal Klopper of McDonnell Douglas Helicopter and Robert N. Salvucci of the General Electric Company.

Authors

Doug Richardson is a defence journalist specialising in the fields of aviation, guided missiles and electronics. After an electronics R&D career encompassing such diverse areas as radar, electronic warfare, and missile trials, he switched to technical journalism. He has been Defence Editor of *Flight International* and Editor of the journals *Military Technology and Economics* and *Defence Materiel*, and his previous work for Salamander includes *An Illustrated Survey of the West's Modern Fighters* (1984) and *Electronic Warfare* (1985).

Lindsay Peacock
Author of the 1992 updated edition, Lindsay Peacock has been a freelance aviation writer and photographer since 1976. He is the author of several hundred magazine articles and a number of books including Salamander's *Strike Aces* (1989).

Credits

Project Manager: Ray Bonds

Editors: Bernard Fitzsimons, Tony Hall

Designer: Robert Mathias, Publishing Workshop. This edition adapted by Studio Gossett

Diagrams: Pete Coote, Mike Keep, Danny Lim, TIGA

Jacket: Stephen Seymour

Colour artwork: Seb Quigley/Bedford Editions, Stephen Seymour

Cutaway drawing: © Pilot Press Ltd.

Filmset: by SX Composing Ltd.

Colour reproduction by Rodney Howe Ltd, Melbourne Graphics Ltd and Scantrans PTE Ltd, Singapore

Printed in Hong Kong

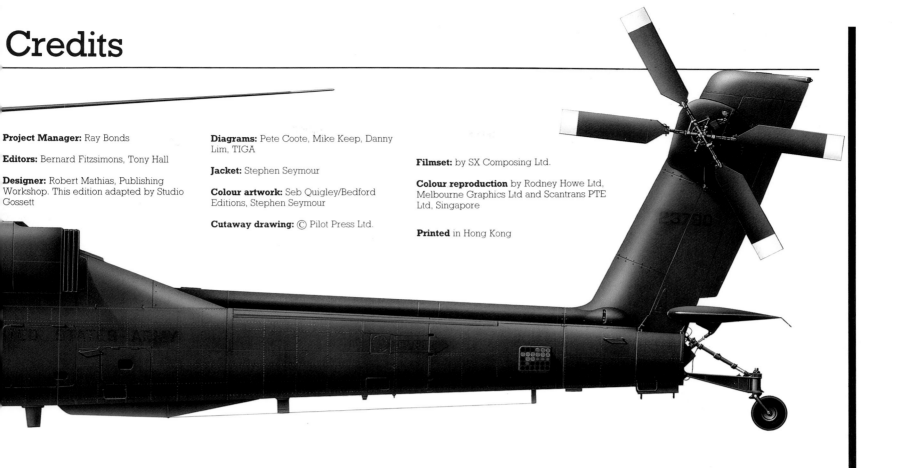

Introduction

In terrain similar to that which formed the hunting grounds of the Native American tribe from which it takes its name, the McDonnell Douglas AH-64 Apache recently had the chance to show that it, too, is a fearless and deadly hunter, when it played such a vital role in the Allied efforts to liberate Kuwait. In just six weeks of combat operations under the code name 'Desert Storm', Apaches of the US Army ranged far and wide by both day and night in the search for targets, and were eventually credited with destroying roughly 1,000 Iraqi armoured vehicles. Perhaps even more remarkable was the fact that this impressive tally was achieved without the loss of a single Apache to enemy action.

Being able to find targets is one thing – destroying them is quite another, but the Apache has several strings to its bow. Foremost among them is the laser-guided Hellfire missile which was used to such deadly effect in the recent Gulf conflict, but it may also employ unguided rockets or an integral Chain Gun. Between them, they enable the AH-64 to engage virtually all of the types of target.

Future Apaches are likely to be even more potent, for both the company and the customer are anxious to improve what is already a star performer. One way in which they hope to achieve this is by the use of radar, and trials of the Longbow Apache have already shown that this should result in enhanced detection and targeting capability in marginal weather conditions. At the same time, both agencies are exploring new weapons options, with a radar-compatible version of the Hellfire missile figuring high on the priority list. While anti-armour operations are to remain Apache's primary task, plans are also in hand to provide an effective defence against aerial threats and these seem certain to result in the addition of the Stinger air-to-air missile.

So, while both McDonnell Douglas and the US Army may derive satisfaction from the Apache's performance in Desert Storm, they are both determined to meet the challenges of the future.

Development

Armed UH-1 'Hueys' and other improvised gunships persuaded the US Army that the attack helicopter was a viable military weapon. The resulting AH-1 HueyCobra played a vital role in Vietnam, backing up the fighting soldiers by providing fast-reacting and hard-hitting firepower. The concept was proven, and the TOW-armed helicopter became an essential component of the US Army strength. The time was now ripe for the attack helicopter to reach maturity in a new design, and despite one false start, the Army ordered the design and fly-off of rival prototypes able to meet the demands of modern combat.

Apache owes its origins to the mass of combat experience with attack helicopters gained by the US Army during the Vietnam War. Although aircraft such as the UH-1 'Huey' and OH-6A Cayuse operated with add-on armament, they were basically transport and scout helicopters respectively. The concept of the attack helicopter was brought to fruition in the Bell AH-1, the US Army's first custom-built attack helicopter: a derivative of the UH-1, it was deployed to Vietnam in August 1967.

These pioneering helicopter gunships played a major role in the Vietnam War, and were generally well liked both by the crews who flew them and by the ground troops who relied on their fire support. In the words of one former Vietnam gunship pilot, 'There's a lot of us walking around that wouldn't be here now if the aircraft hadn't done as well as they did in Vietnam.'

Vietnam combat proved the concept of the helicopter gunship, but the rigours of combat also highlighted deficiencies in the existing aircraft, in particular a lack of engine performance. Operating in tropical conditions, gunships were rarely able to fly with a full weapon load and enough fuel for a useful military mission. Fuel or ammunition had to be removed in order to reduce weight.

The US Army became an enthusiastic attack helicopter operator, convinced of the type's usefulness. Future gunships could operate even in central Europe in the face of Warsaw Pact air defences, argued the Army, provided that sufficient engine power could be installed and vulnerability to ground fire could be decreased.

First attempt to create an advanced attack helicopter was the Advanced Aerial Fire Support System programme of the mid-1960s, which resulted in Lockheed being given a contract in March 1966 to develop ten prototypes of its AH-56A Cheyenne. This was an aircraft very much oriented toward Vietnam-type tactics such as running gunfire at treetop level and diving rocket fire from altitude. The specification concentrated on features such as high airspeeds, suitability for low-level flight and the ability to perform close fire-support missions by flying just above the trees.

The first Cheyenne flew in September 1967, and, as a feat of engineering, the AH-56 was magnificent. Vaguely similar in appearance to the AH-64, it weighed around 17,000lb (7,700kg) when ready for action. Powered by a single 3,435shp General Electric T64-GE-16 turboshaft engine driving a four-bladed main rotor, a conventional tail rotor and an aft-mounted pusher propeller, it could reach a top speed of 220kt (407km/hr). In January 1968 the Department of Defense approved the production of 375 examples.

Unfortunately for Lockheed, the specification to which it was developed began to look unrealistic when the deployment of the SA-7 shoulder-fired anti-aircraft missile in Vietnam made the Army start to question the desirability of flying just above the trees. In future, it was reasoned, attack helicopters would have to operate not above the trees but down among them, so manoeuvrability rather than speed would be the key to survival.

Cheyenne was simply not designed for this type of warfare. Changing requirements, technical difficulties, the loss of a prototype and a conflict between the Army and the USAF over the close-support mission, plus the escalation in threat from medium to high intensity, all conspired to end the Cheyenne programme. Instead of buying the anticipated 500–1,000 AH-56As, the DoD cancelled the programme in the spring of 1969.

In the early 1970s the US Army tried again, drawing up a requirement for an Advanced Attack Helicopter (AAH), an aircraft better suited to nap-of-the-earth operations. The AAH was seen as an aircraft with higher performance, better agility and increased firepower which could supplement and eventually replace the existing Bell AH-1, entering service at the end of the decade. The specification demanded rapid control response for safety in trees, and the ability to manoeuvre around, over or under obstacles such as power lines or bridges; high speed was not requested.

The programme was formally announced in August 1972, the official requirements was approved by the Army in September, and requests for proposals (RFPs) were issued to industry two months later. The requirement placed great emphasis on day, night and adverse weather capability, and on survivability in the front-line battlefield environment of the future. Cruising speed was to be 145kt (269km/hr), and the aircraft was to be able to carry a warload of eight Hughes TOW wire-guided anti-tank missiles for sorties of up to 1.9 hours duration.

Performance was specified not in the traditional sea-level terms, but at 4,000ft (1,220m) altitude with a 95°F ambient temperature, conditions which equate to about 7,000ft (2,135m) plus density/altitude. The latter figure lies in the centre of the 6,000–8,000ft (1,800–2,400m) density/altitude conditions typical of Vietnam combat conditions.

Above left: The Hughes OH-6A Cayuse observation helicopter was too small to form the basis for an attack helicopter, but gave engineers a base on which to draw when creating the YAH-64 design.

Left: The Model 77 mock-up was not too dissimilar to today's Apache, but had a more box-like canopy. Note the armament of TOW missiles and the absence of avionics bays on the fuselage sides.

Left: Lockheed's AH-56 Cheyenne was an impressive feat of engineering, but was developed to a specification which pre-dated the era of shoulder fired SAMs. Its cancellation led to the AAH programme.

Right: Cannon and rocket-armed Bell AH-1G Cobra attack helicopters set out in search of action during the Vietnam War, a conflict that was to prove the viability of specialised rotary-wing attack aircraft.

Below: Flying at the treetop height typical of Vietnam War helicopter operations, an AH-1G demonstrates tactics later made obsolete by shoulder-fired SAMs.

The specification also demanded that the new helicopter should have the manoeuvrability needed for nap-of-the-earth combat flight, including the ability to withstand instantaneous load factors of between +3.5 and −1.5g at maximum gross weight. As if this were not sufficiently demanding, the Army also insisted that the AAH be able to withstand isolated hits from 0.5in (12.7mm) heavy machine guns and keep flying for at least 30 minutes after being hit by a single 23mm cannon shell.

In Vietnam, helos often returned with combat damage. Aircrew were jokingly referred to as 'magnet butts' – characters with a positive talent for attracting ground fire – and when aircraft went down, crew members were often killed or injured. Consequently, when planning its next generation of helicopters – the UTTAS transport and AAH – the Army drew up stringent crashworthiness requirements. The aircraft had to be able to land on a hard surface at a vertical speed of 42ft/sec (12.8m/sec) – approximately 30mph (48km/hr) – with a forward speed of about 15kt and lateral speeds of around 7kt, and offer the crew a 95 per cent chance of surviving.

By early 1973 the US Army had received five proposals. In addition to the established helicopter companies – Bell, Boeing-Vertol, Hughes and Sikorsky – who might be expected to respond with proposals, the AAH requirement also attracted a proposal from Lockheed, which was determined not to abandon the field it had entered with the AH-56A.

Hughes made determined attempts to keep size and weight to a minimum. Engineers looked at the possibility of basing the company's AAH on the OH-6A Cayuse observation helo used by the US Army, but this proved impractical. Next move was to attempt an all-new design of small size, but the ambitious requirements laid down by the Army forced company designers to draw up plans for a helicopter larger and more powerful than any which the company had built to date. This was designated the Hughes Model 77.

Even larger helicopters had borne the Hughes name, but the experimental XH-17 and XV-9A designs, flown in 1952 and 1964 respectively, had been products of the Hughes Aircraft Company, prior to that organisation's bowing out of the aircraft field in favour of electronics and missiles. Hughes Helicopters was an entirely separate part of millionaire Howard Hughes' business empire.

Readers who find the logic of the above difficult to fathom are in good company. In the mid-1970s a cartoon on the wall of the public relations office at Hughes Aircraft showed a puzzled reporter saying to a company spokesman 'Let me get this straight; you're called Hughes Aircraft, but you don't build aircraft – and you haven't seen Mr Hughes for 20 years?'

The original mock-up configuration was not too dissimilar from today's AH-64A. Main contrasts were a different canopy configuration, minimal forward avionics bays, wings fitted for TOW missiles and the absence of the current 'bug-eye' nose turret. In developing the Model 77 Hughes engineers used many of the combat-proven features from the OH-6, including a crushable structure under the crew stations and a static mast.

Revised specification

Before contracts were issued, an important change was made to the specification. The original AAH Material Need document issued in late 1972 stipulated TOW missiles, but studies of the evolving threat persuaded the Army that this missile did not have sufficient range to permit AAH to stand off out of range of many threat systems when launching its weapons.

The newer Rockwell Hellfire was still at an early stage in development, but offered a range of at least 3.7 miles (6km), while the use of a laser seeker gave the prospect of fire-and-forget attacks. The gamble of switching to an unproven missile seemed worthwhile, so in early 1973 Hellfire replaced TOW as the planned AAH armament.

This switch in missiles had a significant impact on aircraft weight and cost. Mission requirement for the existing TOW-armed AH-1 Cobra was 1.9 hours, so the initial requirement for the TOW-armed AAH had specified 1.9 hours, but the use of Hellfire involved additional mission equipment, making the aircraft heavier, so the Army decided to reduce fuel load – and thus endurance – in order to maintain aircraft weight.

On June 22, 1973, the DoD announced that the Bell and Hughes designs had been selected for competitive development, and contracts were awarded to both companies for Phase I of the AAH programme. Bell received $44.7 million to develop, build, and test-fly its design under the designation YAH-63, while Hughes received $70.3 million for a similar programme involving its Model 77, to be known as the YAH-64.

Both contracts covered the design of the aircraft, plus the construction of a static test airframe, a ground test vehicle (GTV), and two flying prototypes, and first flight of both prototypes was scheduled for March 1975. Following a seven-month Government test and evaluation programme, the winner was to be announced in July 1976.

The designs which emerged from the rival companies were similar in general configuration to the Army's existing AH-1, but much bigger and heavier. The

Right: Unsuccessful AAH contender was the Bell YAH-63. This design featured twin-bladed main and tail rotors, a tricycle undercarriage, and a cockpit which placed the pilot ahead of the gunner.

AH-1 weighed around 6,500lb (2,900kg) ready for action, but the new Bell and Hughes designs would weigh around 13,000lb (6,000kg). Both were twin-engined, and used the same powerplant, the General Electric YT700 powerplant already being developed for the Army's UH-60 UTTAS helicopters.

A nominal look was taken at alternative powerplants to the T700. The Army had not formally ordered the use of the GE engine, but the use of this powerplant was almost implicit. In theory, Bell and Hughes could have proposed a different engine, but they might have a hard time persuading the customer to accept it.

The most obvious external differences between the two designs were in fuselage shape and number of rotor blades. Bell's fuselage was rounded in cross-section, tapering away to terminate in the tail boom. The vertical tail surface extended above and below the tail boom, giving the aircraft a shark-like appearance. The twin T700 engines were located in housings mounted below and behind the cockpit.

All helicopter designers face the task of converting the high-speed shaft rotation from a turboshaft engine to the much slower speed required for the rotor. For the YAH-63, Bell designers decided to connect a drive shaft directly to the engine output. Turning at high speed, this would transfer energy to the transmission, which was designed to provide the full degree of step-down required to power the rotor system.

The main rotor was of the twin-blade configuration proven in combat by the fleets of UH-1 and AH-1 helicopters used in Vietnam. One minor disadvantage was its 51ft (15.5m) diameter – 3ft (91cm) greater in radius than the four-bladed unit used by Hughes on the YAH-64. The twin-bladed tail rotor was located on the port side of the tail and was driven by an externally mounted shaft which ran above the upper surface of the tail boom, giving good access for maintenance.

Tadpole style

If the Bell designers had created a flying shark, the Hughes designers seemed to have taken their inspiration from the tadpole, coming up with a more angular-looking design with a slab-sided fuselage and a more prominent transition between main fuselage and tail boom. The housings for the T700 turboshafts were located higher on the fuselage, and slightly further to the rear than on the YAH-63.

YAH-64 engineers decided to tackle the problem of linking the engine and transmission in two stages, fitting a gearbox (known as the nose gearbox) close to the engine in order to step down the speed before applying energy to a drive shaft which would be used to transfer torque to the transmission.

One prominent feature which the YAH-64 inherited from the OH-6 was the use of a four-bladed main rotor. Discussing the relative merits of two-bladed and three/four-bladed rotors with the author, an Apache engineer admitted that 'It's a little bit like religion and politics – you can get opinions and argue about them all day and not reach an absolute conclusion'. The use of three- or four-bladed main rotors had been a Hughes tradition. For two helicopters of similar weight, the smaller blades of a three/four-blade design carry smaller loads, the company argued, and increase the frequency of the resulting vibration.

Smaller blades also result in the rotor sweeping out a smaller radius, providing significant advantages in low-level combat. Experience in Vietnam had shown that during combat operations down at low level amongst trees, the OH-6 could operate in more densely wooded locations than the UH-1 or OH-58 thanks to its smaller rotor.

The tail rotor was also of four-bladed configuration. Located on the port side of the vertical tail, it was driven by a shaft mounted within a spine running along the top of the tail boom. The top of the YAH-63 fin carried a small horizontal surface, but the Hughes team opted for a much larger stabiliser positioned at the base of the vertical fin. Bell chose to locate the pilot in the front cockpit, the gunner above and behind him, while in the Hughes aircraft, this arrangement was reversed.

Combat experience with the AH-1 had shown the virtue of a turret-mounted cannon. The AH-1S had carried a 20mm M197 steerable cannon, but for the AAH the Army wanted a more powerful gun, specifying a turret-mounted 30mm weapon plus 800 rounds of ammunition. The most obvious source of hardware to meet the new requirement was General Electric, whose Burlington plant had supplied many US airborne guns, including the M61A1 Vulcan fitted to most US jet fighters, and gun designers at GE's Burlington facility in New England offered the XM188, a three-barrelled Gatling-type weapon.

GE was not the only contender, however. Hughes Helicopter had its own ideas on cannon design, and since late in 1972 company engineers had been working on a weapon using a novel operating system – the XM230 Chain Gun. Early in 1985 both gun manufacturers were given contracts to deliver two weapons of each type to the US Army for competitive evaluation.

Above: Months of burning the midnight oil allowed Hughes engineers to catch up with and eventually overtake their Bell counterparts, flying YAH-64 AV02 on September 30, 1975.

In positioning the electro-optical sensors and the cannon, the two teams again took different approaches. Bell engineers decided to fit the General Electric XM188 cannon on the underside of the nose, while the electro-optical sensors were mounted in a trainable belly turret directly below the front of the pilot's windscreen. Hughes reverses the undernose location for the gunsight and laser, positioning its XM230 Chain Gun directly beneath the front cockpit of its aircraft.

AAH was obviously going to be more expensive than the AH-1. In order to keep costs down, the Army inserted clauses within the contracts which

Above: The stepped cockpit gives the pilot a good forward view over the gunner's head. The tiny chin-mounted sensor turret was fitted to both YAH-64 prototypes, and was intended for trials purposes only.

Right: Bell's YAH-63 was the testbed for the General Electric XM188E1 cannon, a three-barrelled Gatling weapon which was to be rejected as the AAH armament in favour of the revolutionary Hughes Chain Gun.

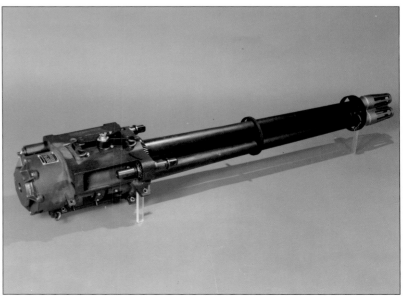

allowed compromises to be made in the event that these would result in significant savings in cost or timescale. Unit fly-away cost of the aircraft was to be $1.6 million in 1972 dollars, and the entire cost of developing the aircraft and manufacturing a planned production run of 472 examples was estimated at just over $3 billion (also in 1972 dollars).

As the prototypes took shape, it became obvious that both companies were spending money at a higher rate than had been anticipated, and to keep costs within limits a six-month slippage of the programme was agreed in early 1975. In addition to delaying the programme, this also penalised the competitors: both lost the incentive payments which the Army would have made had the programme remained on time.

When the YAH-64 design was first drawn up, a low position for the vertical tail seemed the most obvious solution. This would protect the tail rotor when settling the aircraft down into the trees, whereas a T-tail configuration would have imposed structural complications as a result of the need to support a large horizontal surface at the top of a relatively small vertical surface.

This original tail design never flew on a YAH-64. Flight trials using a modified OH-6 fitted with a wooden low-set tail soon highlighted potential handling

problems. As the aircraft transitioned from the hover into forward flight the airflow from the main rotor blew back over the tail, creating pitch moment changes which pilots found dificult to compensate for. A significant pitch trim change was also needed when transitioning from level flight to a maximum-power climb.

Although the tail boom of the prototype had all the fittings for the low-set tail, the simplest cure for the handling problems seemed to be a T-tail, accepting the resulting structural complications as the price to be paid for improved handling during transition. The revised tailplane ready in time for the first flight.

On April 19, 1975, Bell started ground running, proof-load, and vibration testing on the YAH-63 Ground Test Vehicle (GTV), and two months later Hughes began running its GTV. Faced with Bell's lead in timescale, Hughes poured man-hours into an effort to catch up, with engineers working long hours to gather data from the ground rig and prepare the first prototype for flight. In general, the task went smoothly on the Hughes side. Few problems emerged, and the massive expenditure of midnight oil allowed the company to fly prototype AV02 (Air Vehicle No 2) on September 30, 1975, at Palomar Airport in Southern California. There was no AV01 – the first airframe had been the GTV.

The first Bell YAH-63 flew on the following day. One factor which may have slowed Bell was the high-speed drive between the engines and the transmission. The price paid for eliminating any intermediate gearbox was problems with these highly stressed components. November 22 saw AV03, the second Hughes prototype, take to the air, while Bell flew its second aircraft for the first time on December 21, 1975.

Contractor testing continued through the first half of 1976, with all four AAH

Above: Prototypes AV04, 05 and 06 pose for the camera. Note the absence of a horizontal stabiliser on AV05, a configuration briefly tested during the search for a definitive stabiliser position.

prototypes scheduled for delivery to Edwards AFB in the summer. March of that year saw the XM203E1 Chain Gun and 2.75in rockets fired from the air for the first time, and the rival patterns of cannon beginning three months of ground tests at Rock Island Arsenal.

On June 4, just before the scheduled hand-over, one of the YAH-63 prototypes was damaged in an accident while being flown by Army pilot in the front seat and a Bell test pilot in the rear. An investigation later showed that the tail rotor drive shaft had failed while the aircraft was in sideways flight to the right. This is a high-load condition, involving the use of a lot of left pedal and pitch on the tail blades. The drive shaft failed, and without the anti-torque effect of a powered tail rotor the aircraft made just over two turns, struck the ground and flipped over. Both crew members survived the accident, but the aircraft was a write-off. In order to provide the Army with a second flying prototype, Bell had to upgrade its GTV to flight standard.

Once delivered to Edwards AFB, both AAH designs were put through a competitive fly-off, and while this was under way another part of the AAH was taking shape – the complex and sophisticated sensor suite whose two main items were the Target Acquisition Designation Sight (TADS) and the Pilot's Night-Vision Sensor (PNVS). TADS, to be used by the gunner, combined direct-view optics with a forward-looking infra-red sensor, TV system, laser rangefinder/designator and a laser tracker. The two systems were designed to work together, with sensors mounted side-by-side in a single

Top: AV02 and 03 could never be confused with production aircraft. Recognition points include the T tail, short avionics fairings on the forward fuselage, and the tiny chin-mounted sensor turret.

stabilised turret. As a result, the installation is usually referred to as the TADS/PNVS. On November 27, 1976, industry submitted its TADS/PNVS proposals which, like the airframe and cannon, were to be the subject of competitive development.

On December 10, 1976, the Army announced a double success for Hughes. The winning AAH design was the YAH-64, and the Chain Gun was selected as its cannon armament. Factors which swung the decision away from the Bell aircraft included its two-bladed rotor and the small triangular footprint of its tricycle landing gear. On uneven terrain the latter would have made the Bell aircraft easier to roll over accidentally.

Programme Phase II

Hughes Helicopters now faced Phase II of the AAH programme, an effort tasked with clearing the aircraft and its systems for full-scale production. This was to have lasted for 50 months, but the US had just seen a change of Government, with incoming President Jimmy Carter taking over from Gerald Ford, and the two Administrations had markedly different views of the priority which should be given to AAH. The Ford Administration had earmarked $200 million for AAH Phase II development in FY78, but Carter decided to halve this.

This cutback had an immediate effect on the programme. Hughes was forced to fit its programme within the shrunken budget, and drew up a scheme under which three more prototypes would be built and fitted with weapon systems, and the two existing aircraft would be

Above: Problems with the horizontal tail were to dog the development programme. Here one of the original YAH-64 prototypes assesses the effects of vertical surfaces at either end of the stabiliser.

updated so that they too could be used for weapon system trials. Instead of lasting for the planned 50 months, this 'lean' version of the Phase II programme would run for 60.

Congress disagreed with the magnitude of Carter's cutback and authorised a compromise figure of $165 million. This partial restoration allowed Hughes to trim four months off the Phase II schedule, and the contract awarded to the company was finally settled at 56 months' duration, and a total value of around $390 million.

During 1977 a package of modifications known as Mod 1 was gradually applied to the two flying prototypes and the GTV. Initial flight trials with AV02 and AV03 had highlighted two problems. When flying simulated nap-of-the-earth missions both aircraft tended to fly in a slightly nose-up attitude, degrading forward vision and presenting the pilot with the need to lower the nose using the flight controls before missiles or rockets could be fired. On other flight tests which subjected the aircraft to extreme flight conditions, the main rotor blades had actually struck the top of the canopy.

The rotor problem was easily fixed – Hughes engineers redesigned the mast, increasing its height by 9.5in (24cm) – but the attitude problem proved less tractable. Both prototypes had been fitted with a fixed T-tail assembly whose rigidly mounted horizontal stabiliser was located at the top of the tailplane.

The dynamic environment of helicopter operation exposed the stabiliser to air loads and vibrations, the very problems which had originally persuaded

Above: AV04 was the first prototype to fly with the low-set moving stabilator, and a tail rotor located further up the vertical fin. The IR suppressors fitted on the engine exhausts had already been flown.

Right: Apache pays a visit to Ryan Aeronautical, the company which manufactures most of the aircraft's structure. The TADS/PNVS turret is one of the wooden mock-ups often fitted to AH-64 prototypes.

Hughes to avoid the use of a T-tail if possible, and it soon became clear that an alternative design of stabiliser would be needed. AV03 was flight tested with a low-set stabiliser – the configuration originally rejected prior to the first flight – while at one stage Hughes even flew an Apache over a limited range of speeds with no horizontal stabiliser fitted. The revised stabiliser fitted as part of the Mod 1 scheme was lighter in weight than the original, had a straight leading edge and swept trailing edge (on the original component the leading edge was swept and the trailing edge was straight) and, like the earlier pattern, was fitted at the top of the vertical fin.

Changes were also made to the tail rotor, a component which plays a major role in ensuring good controllability in sideward flight. In the past many design teams have erred in making the tail rotor too small, with the result that a point can be reached in sideward flight where the aircraft becomes difficult to handle. Given the fact that Apache was destined to spend its life down among the trees, an environment where good lateral control would be needed, the designers decided to create an improvement in tail rotor authority, increasing its diameter by 3in (7.6cm).

Compressibility effects had also been experienced with the main rotor. As a helicopter rotor revolves, the loads imposed on each individual blade vary throughout the 360°, and trials with Apache showed that a large peak load was being experienced on the advancing side of the rotor (that is, on the starboard side of the aircraft). Passing into the blade-retention system and airframe, this was a potential source of fatigue problems. A peak in sound level was also experienced. Sweeping the blade tip softened both the load 'spike' and the sound impulse. A further 6in (15cm) extension of the main rotor mast was also carried out.

Another change involved the replacement of the engine-driven fans used to reduce the IR signature of the exhausts by passively-cooled exhaust ducts known as Black Holes, and flight tests at White Sands have confirmed the aircraft's low IR signature.

On March 10, 1977, exactly three months after the announcement that Hughes had won the AAH competition, contracts for TADS/PNVS development were awarded to two rival teams. Funding worth $55 million was split between the chosen candidates, Martin Marietta and Northrop, and following a fly-off of the rival systems aboard AH-64 full-system prototypes the Army hoped to choose the winner in February 1980.

As was the case with the airframe, the two teams took different approaches, Northrop planning to install the PNVS in a cavity above the helo nose and the TADS underneath; unless required, the PNVS would be retracted. Martin Marietta favoured a non-retractable installation.

This development work continued, but Apache's future began to look insecure. AH-64 was a dramatic step forward for the Army in terms of attack helicopter capability, but complexity inevitably drives up cost, making Apache the most expensive helicopter

Above: A fly-off was carried out in 1980 between the rival Martin Marietta (left) and Northrop (right) patterns of TADS/PNVS. In April of that year the Martin Marietta design was chosen for US Army use.

Right: AV04 – seen here with stabilator at near-maximum deflection – flew many sorties intended to prove tailplane design. On November 20, 1980, it crashed after colliding with its chase plane.

ever developed by the USA and the inevitable target of criticism. Much of the cost escalation due to improvements and the effect of the vicious inflation which followed the 1973 Middle East War was little reported at the time. When Congress suddenly became aware of the revised figure, the result was bound to be opposition.

Memories of the AH-56 Cheyenne helped focus opposition against another complex and expensive helicopter, while the ending of the Vietnam War blunted perception of the evolving threat which AAH was designed to face, and many critics drew ammunition from a report which had been prepared by the General Accounting Office (GAO) of the US Congress. This questioned the likely effectiveness of the aircraft and its missile armament when faced with the high-intensity type of combat operations envisaged for the future.

In its final form, the FY78 defence budget contained only half the requested Engineering Development funding, and the Army was ordered to review its requirements. In retrospect, the programme was lucky to avoid outright cancellation.

US Government Engineering Design Tests were completed in April 1978. By May 1978, all the Mod 1 engineering changes had been flight tested, and both aircraft were grounded in order to be reworked with a further series of modifications. Known as Mod 2, this would bring the prototypes closer to the planned production configuration. The most obvious external change was a marked increase in the size of the equipment fairings on the fuselage sides.

In the original configuration, these extended rearward from just aft of the nose, ending directly beneath the point where the canopy divided between front and rear sections. Under Mod 2, the two aircraft were to be equipped with armament, fire-control and navaids, and in order to provide the additional space needed by these new avionics units, the fairings were extended rearward, ending just below the leading edge of the stub wing.

The flat transparent panels of the canopy had been found to vibrate in flight, a phenomenon known as drumming. This vibration was apparently induced by the rotor, so the canopy was redesigned to use panels which were slightly curved in one plane only for improved stiffness.

During Phase 1 testing AV02 and AV03 had flown for around 700 hours in total, and both were now fitted out for further test flying. AV02, assigned to the task of proving the aircraft structure, restarted flight testing on November 28, 1978. Instrumented with strain gauges, it was used to check the structure and rotor throughout the flight envelope. AV03 was used for propulsion and flying qualities testing, returning to flying duties on December 29. By the autumn of 1979 the two aircraft had flown for a total of 1,000 hours, but despite the revised tail, the attitude problem had still not been settled. As originally predicted, the T-tail configuration imposed unacceptable stresses on the rear fuselage, and would have limited airframe life to less than the 4,500 hours that had been specified by the Army. In order to get out of the dynamic and load problems experienced with the T-tail, and to achieve a level attitude of the aircraft under a wide range of flight conditions ranging from 25 to 30mph (40-48km/hr) in rearward flight to 200mph (320km/hr) in forward flight, engineers were forced to accept the complications of adopting a moving stabilator.

Revised tail

AV04, first of the second batch of prototype aircraft, was fitted with a revised tail assembly. The horizontal stabiliser remained in the low-set position first tested on AV03, but was now of variable incidence, while the vertical tail surface was increased in height by 3in (7.6cm) and the tail rotor was positioned 30in (76cm) higher to maintain clearance with the definitive design of stabilator.

AV04 was originally due to fly in June 1979, but by the late spring of that year delays in the delivery of components from subcontractors had caused the date to slip to August. The programme continued to slip through the summer and autumn; AV04 finally flew in November 1979, joining the fleet as a structural and performance testing aircraft.

Several sizes of stabilator were flown on AV04, but the final and definitive version, slightly smaller than the original, was flown on March 1980. AV05 flew in December 1979 and was used for system testing. AV06, the final prototype of the second three-ship order, was not flown

Right: Apache manages to look menacing with even a half-load of Hellfires. The presence of a red nose probe shows that this was a development test flight.

Development

Above: Prototype AV02 was the last to abandon the original T-tail configuration. One of its main tasks was to test the Apache armament and fire-control system. It is seen here firing 2.75in unguided rockets.

until March 1980. AV06 was the first aircraft to combine this stabilator with a tail rotor 10in (25cm) larger in diameter than the earlier model.

Before being flown, Apache systems were integrated and tested in the laboratory. After being checked out in bench tests the individual systems were interconnected, then put through further bench tests. By the late spring of 1979 most of the Apache systems were over this hurdle and installed in the mission systems simulator. This took the form of a partial fuselage, including the crew stations and all the aircraft's wiring harnesses, and allowed pilots and gunners to 'fly' the aircraft as its systems were exercised. It also served as a useful simulator which could be used to train company and Army aircrew prior to the start of system flight tests at the Yuma Proving Ground.

These flight tests included a fly-off between the rival TADS/PNVS systems, prototypes of which began test flying on YAH-64 prototypes in the summer of 1979. The Martin Marietta system was installed in AV02, the Northrop equipment in AV03, and January 1980 saw the start of the Army's fly-off. This lasted for two months, and included day and night firings of Hellfire missiles against targets designated both from the ground and autonomously using the TADS in the helicopter.

Selection of the Martin Marietta equipment was announced in April 1980. The equipment had done well during the trials, scoring three Hellfire hits in three firings, and the company was given a $45.8 million contract for a 26-month programme of work on the system. This covered further development intended to prepare the equipment for start of production in December 1981. It also contained options for initial production quantities.

By the autumn of 1980 many Apache flight-test goals had been achieved as the five-strong fleet tackled various engineering assignments. After completing a scheduled modification programme, in which it was given the new-pattern stabilator and tail rotor, AV03 flew a 170-mile (273km) ferry mission – the longest non-stop journey so far attempted by a YAH-64 – then flew a series of handling trials.

AV02 had completed a series of armament and fire-control tests during which it scored a Hellfire missile hit on a tank target at the longest range so far achieved. AV02 had been the last aircraft with an old-pattern stabilator and tail rotor, but this was put right in a rebuild programme. It was used for further armament trials, including firings of Hellfire, Chain Gun and 2.75in rockets.

Clearance of the full flight envelope came that autumn when AV04 reached a speed of 206kt (381km/hr). During more than 1,500 hours of flight time, Apache prototypes had also flown manoeuvres of more than 3g at speeds of 80–164kt (148–304km/hr). The automatic stabiliser was performing as predicted, while the new tail rotor had allowed speeds of up to 45kt (83km/hr) in right sideward flight. It was then used in firing trials intended to confirm that weapon firing did not threaten the stabilator with excess stress or damage due to debris. AV05 was engaged on propulsion and power-system trials, while AV06 shared the task of armament testing with AV02.

On November 20, 1980, AV04 was despatched on a sortie tasked with in-

Left: Prototype AV02 – still equipped with the high-set T tail – was also used to launch development rounds of the Hellfire anti-tank missile. On trials, hits were scored at ranges which delighted the US Army.

Above: An anti-Apache faction in Congress was always on the lookout for cheaper alternatives to the Hughes aircraft. One contender was the Hellfire-armed version of the Sikorsky UH-60 Blackhawk seen here.

Right: What seems at first sight to be a new camouflage scheme is in fact the result of airframe icing. Those yellow patches show where ice formed on the structure during a trial of the Apache de-icing system.

vestigating the effect of tail incidence on drag. For this mission it was accompanied by a T-28 photographic chase aircraft which was to record the behaviour of a tufted tail. As the two flew in close formation so that a photographer in the rear cockpit of the T-28 could observe the helicopter tail, the two aircraft collided and crashed. The sole survivor was the pilot of the T-28.

As delays and minor technical difficulties forced up the cost of the programme, AAH again came under criticism, and the threat of cancellation loomed once more when the Pentagon decided to eliminate the programme from the FY81 defence budget. Since the Army had developed the AH-1 by arming the existing UH-1, moves to question the need for an all-new AH-64 were predictable, given the fact that the new Sikorsky UH-60 Blackhawk could in turn act as a starting point for an AH-1-style derivative aircraft. The Army was finally able to have funding restored, but the DoD ordered the Service to carry out a comparative evaluation of the AH-64 and a Hellfire-armed UH-60.

In February 1981 Hughes received $25.1 million for long lead time items for the first planned production batch, and shortly afterward the company turned over three prototype aircraft to the US Army. These were the original two prototypes, AV02 and 03, plus AV06, and training service aircrew and ground crews on these aircraft started in May 1981. Pilots and ground crew found transition to Apache easy: within weeks Army pilots and gunners were flying

their new mounts in preparation for three months of planned Operational Test II trials. (The first phase of service operational testing had been the fly-off between the rival prototypes at Edwards AFB back in 1976.)

By the summer of 1981 the five prototypes had clocked up around 2,500 flying hours, 1,000 of which were logged by the second prototype. Development was by this time virtually complete, announced Project Manager Maj. Gen. Edward M. Browne, and integration of the complex weapon and avionics systems was on schedule.

Operational Test II was carried out summer and early autumn of 1981, and was intended to assess the operational effectiveness of the aircraft and its weapons in a tactical environment, to show that the AH-64 met the US Army's targets for reliability and maintainability, and to evaluate training procedures.

This was to be no paper exercise – during three months of test flying all elements of the weapon system were tested under simulated battlefield conditions. The aircraft were exposed to temperatures of more than 110°F (43°C), and to fine dust, while soldiers from the US Army's 7th Infantry Division took the part of supporting infantry and hostile forces in order to make the tests as realistic as possible. This gruelling evaluation was completed in August 1981, by which time the three trials aircraft had flown more than 400 hours in three months. This exercise helped increase the total number of hours flown by Apache, and the 3,000th flying hour in the programme was logged during August. Final production go-ahead was withheld until the results had been fully studied. This involved postponing DSARC III from December 1981 to March 1982.

Even before the Army and DoD had given the go-ahead for production, Hughes took the gamble of starting work on the necessary production facilities. In July 1981 the company announced that it would build a new 240,000sq ft (22,300m^2) production plant at Mesa, near Phoenix, Arizona. Planned as the base for final assembly and acceptance flight tests, the $20 million plant would eventually employ 1,800 workers, some transferred from Culer City to the new site, but most being locally hired.

The schedule was ambitious, driven by the need to get production tooling in place and the assembly line running in time to meet the scheduled delivery date for the first production aircraft of November 1983. Ground would be broken before the end of 1981, and hiring of personnel would begin in the second half of the following year.

Full-scale production was finally ordered in April 1982, by which time other test goals had been attained: by January 1982, demonstrations of the armament and fire control systems had been completed, while March saw the completion of cold-weather testing.

Funding withheld

The Army had requested $365 million for procurement of the first 14 AH-64 production helicopters under the FY82 budget. While this figure was eventually increased to $444.4 million by Congress, it provided for only 11 aircraft. The US Senate Armed Services Committee refused funding for an initial batch of 48 aircraft until a final agreement on costs could be worked out and the Committee could be given further information on the requirement which AH-64 would meet.

Until early 1981 all development flying had been carried out using the T700-700 engine, powerplant of the UH-60 Blackhawk, but on January 15, 1982, AV05 began test-flying with the definitive Apache powerplant, the T700-701, a land-based version of the uprated -401 engine used in the US Navy's SH-60 Seahawk. Qualification testing ended in May. The aircraft was then used to gather flight data required for the pilot's flight manual, before being assigned to the task of flight-testing the composite main rotor blades planned for production aircraft. These were being developed under a US Army Manufacturing Methods & Technology contract.

AV03 spent the early part of 1982 at Minneapolis in Minnesota, carrying out icing and cold weather tests, then toured US military bases during the summer. AV06 flight tested an explosion suppression feature of the Apache fuel system, then was used as a testbed for the production version of the TADS/PNVS.

In July 1982 Apache set off on its first European tour. In the course of its 5,700-mile (9,170km) trip, AV02 was demonstrated to European government and military officials, and to the US forces in Germany who would have to deploy the type. The trip reached a climax in September with the aircraft's appearance at the Farnborough 82 and Army Air 82 shows in England. By the time the aircraft returned to the USA and headed for the Association of the US Army's annual meeting in October, it had accumulated 154 flight hours.

While Apache was impressing the Europeans, Hughes submitted its proposals for the second production batch of aircraft, Lot 2, which was to consist of 48 aircraft. The Army had increased its planned production run from 472 to 536 before being forced to trim this back to 446 in an effort to reduce programme

Right: Rebuilt to production configuration, including the definitive low-set stabilator and revised vertical fin, the hard-worked prototype AV02 was one of the star attractions at the 1982 Farnborough Air Show.

costs, but Hughes estimates indicated a total programme cost of $5,994 million, which was somewhat higher than the Army's estimates.

In some circles, there was no doubt at all as to the reason for the cost escalation. 'The Pentagon's plans to purchase these helicopters are being subjected to serious criticism in Congress because of their great cost, which in the past year alone has risen by 50 per cent and has now reached $13 million for each helicopter', Moscow Radio broadcasters Yevgeniy Kachanov and Vitaliy Sobolev told their listeners in March 1982. 'The American military . . . say it is a question of US security, of the future of the US Army, and concern, of course, about the Soviet military threat too. But of course, it is a question of something else entirely . . . it is the corporations that make them above all who benefit from the development of expensive types of weapons – the monopolies closely connected with the Pentagon, which defends their interests'.

Despite the problems with cost, Apache's supporters were making their views felt. One concept which had to be dashed was the idea that a low-cost alternative could be fielded at a unit cost of around $4 million rather than the $15.1 million programme unit cost then applicable to Apache. Some critics believed that a low-cost AAH could be created by improving the existing Bell AH-1 Hueycobra.

'The AH-64 is by far the best current solution', Army Undersecretary James R. Ambrose told the House Armed Services Committee on March 26. 'There is no $4 million attack helicopter in the real world.' The best modified version of the Cobra which might be developed was that proposed to meet the West German PAH-2 requirement. This would take four years to develop at a cost of $297 million, he told the Committee. The PAH-2 Cobra would be slightly more expensive than Apache, and could not match the Hughes aircraft's survivability, crashworthiness, or agility. 'The Cobra airframe designed in the 1960s with a single engine simply cannot compete with the AH-64 in these areas, and the PAH-2 version has four times the vulnerable area exposed to threat weapons.'

AAH requirement

In a letter dated July 22, 1982, General Bernard W. Rogers, NATO Commander-in-Chief Europe, spelled out to Senators Ted Stevens and John Tower and Representatives Joseph Addabbo and Melvin Price the need for Apache. 'Since the early 1970s I have been involved personally in the Army effort to field an AAH, and have seen previous attempts fail. During this same period, the Soviets have made steady and dramatic advances in this field where we wrote the book . . . We need the AH-64 in Europe now, and cannot afford the luxury of another trip to the drawing board.'

Successful renegotiation of the Hughes contract resulted in a production go-ahead for Lot 2 in November 1982, Hughes receiving three contracts totalling $105.6 million for the initial funding of the next 48 aircraft; US Army plans now assumed an eventual total of 515 aircraft.

While politicians and accountants haggled, the new plant at Mesa continued to take shape. Ground had been broken back in March – the original schedule of late 1981 had proven over-optimistic – and even before the plant

was ready for occupation pre-production activity was under way in a leased facility at the nearby town of Tempe. By the autumn of 1982 the floors of the Mesa plant had been laid, and the roof of the assembly building was in place. In swarmed the Apache team, and manufacturing work started in December, two months ahead of schedule.

By February 1983 the Apache prototypes had amassed 4,000 combined flying hours, and suppliers around the USA were delivering components and sub-assemblies to Mesa. By March the Mesa assembly line was ready to begin assembly of the first production aircraft, only a year after building work started.

Like most US military aircraft, Apache is the joint product of many countries. Hughes made no attempt to hog as much of the work as possible, but shared it with companies in 36 states in the USA, as well as contractors in Canada and West Germany. One of the biggest subcontractors is Teledyne Ryan Aeronautical, who supply most of the structure – the fuselage, empennage, wings, engine nacelles, avionics bay and canopy. The first shipset was completed at the company's San Diego works in late March, then trucked to Mesa.

Before leaving Teledyne Ryan, the fuselage was mated to an ATAF (Apache Transporter/Assembly Fixture) developed by Tracor, which supported the assembly during the journey. Fuselage and ATAF arrived at Mesa on March 28, but they were not due to part company for some time. Watched by an audience of company employees and local dignitaries, the ATAF and its cargo were installed on Assembly Station 1.

Above: Having proved the new tail design, the ill-fated AV04 was able to give a brief glimpse of how the service aircraft would fly and fight at low level. Today's Apache crews fly even closer to the ground.

At this point, ATAF demonstrated its second use as a flexible fixture which permits the fuselage to be elevated to a suitable height for assembly work and rotated by up to 90° in order to ease access, and within minutes technicians had closed in to begin the task of turning this empty shell into a production Apache. The line was rolling and Production Vehicle 01 (PV01) was taking shape. The goal was Assembly Station 22, followed by a rollout in September.

By May 9 PV01 had reached Assembly Station 5, and throughout that

Right: September 30, 1983, saw the rollout ceremony for the first production Apache, but that red instrument boom was a clear indication that PV01 would be staying at Mesa for flight-test duties.

month, as work continued on the aircraft, construction of the Mesa plant pressed ahead. The warehouse, paint shop and flight hangar neared completion, while work took place to convert a nine-acre site into flight ramps and to dig a 25ft (7.6m) deep bowl-shaped pit on another two-acre site. The latter was due to house a whirl cage intended to reduce the nose of rotor blade testing and other work involving the Apache GTV.

The fuselage for PV02 arrived at Mesa in May, a month which also saw Apache head over to Europe for a second time and an appearance at the Paris Air Show. The planned production programme would require significant numbers of test pilots, so mid-1983 also saw AV03 being used to train company test pilots and US Army acceptance pilots. Training started at Yuma, then moved to Mesa.

The third fuselage reached Mesa in July, and by the 22nd of that month – the date on which completion of the Mesa Assembly and Fight Test Centre was announced – it was already at Assembly Station 2. By then, PV01 had worked its way down to Station 11 at the far end of the assembly building and had begun the journey back along the other side of the facility. It was now at Station 14,

Below: First flight of PV01 took place in front of a backdrop of earth mounds – temporary end products of a fast-moving programme which had seen the new Mesa assembly building erected in only six months.

directly opposite Station 9 where PV02 was being assembled. In August, Congress authorised the FY84 Apache buy – a batch of 112 aircraft.

By the end of September PV01 had finally completed its journey to Assembly Station 22. Watched by an audience of 1,600 guests and accompanied by an Apache Indian on horseback and a US Army ROTC colour guard, it was rolled out on September 30, two months ahead of schedule. Five more aircraft were visible on the assembly line, and more than 900 personnel were now employed at Mesa, more than twice the number on strength at the beginning of the year but less than half the total needed to meet the mid-1980s production rate of 12 aircraft per month.

Following the ceremony PV01 began an extensive series of ground tests intended to check out its sub-systems, and by November 17 it had been joined by PV02. The first production aircraft to carry full mission equipment, the latter had been completed six weeks ahead of schedule. Another six were now at various stages of assembly.

Hughes for sale
While this work was going on, time was running out for Hughes Helicopters. Following the death of Howard Hughes in 1976, the industrial empire he had built faced dismemberment, and in July 1983 the chairman of Hughes Helicopters – a trustee of the Hughes estate – announced that he planned to sell the company for more than $500 million.

A purchaser soon emerged in the form of US aerospace giant the McDonnell Douglas Corporation. On December 16, 1983, the two organisations reached a purchasing agreement, three weeks later the deal was completed, and on Friday January 6, 1984, McDonnell Douglas announced that it had acquired Hughes Helicopters for $470 million. Hughes Helicopters initially operated under its own name as a subsidiary of McDonnell Douglas.

Following the take-over, McDonnell Douglas decided to end the production of aircraft at Culver City, cutting the workforce at that location from around 5,000 to some 1,800 and opening a new Hughes 500 plant at Mesa, while the slimmed-down Culver City plant would concentrate on the production of castings, forgings, sheet-metal assemblies, and the Chain Gun.

PV01 flew for the first time on January 9, 1984, and the 30-minute sortie confirmed the end result of more than 4,500 hours of test flying with the prototypes. Hughes chief test pilot Steve Hanvey stated afterward that PV01 had flown smoothly and responded crisply to pilot inputs – just like the prototypes, if not better. Two weeks later, on January 26, it was formally accepted by the US Army when Hughes executives handed over the log book to the commanding officer of the US Army representatives based at Mesa.

The handover was a mere formality: like PV02, the aircraft would never see service with an Army unit. Heavily instrumented and fitted with telemetry equipment, it carried more than 1,000lb (450kg) of test equipment, and both aircraft would remain at the Hughes plant for the duration of the AH-64A programme, where they would be used as trials aircraft.

In February 1984 the successful development and production of Apache

resulted in US Army and Hughes winning the Collier trophy, an award presented annually by the National Aeronautic Association for 'the greatest achievement in aeronautics and astronautics in America.'

In the same month a former Apache pilot carried out a feat of flying well beyond the capability of his previous mount. Col Robert L. Stewart had been a senior Apache test pilot back in 1976 during the YAH-63/-64 fly-off. Now assigned to the Space Shuttle programme as the Army's first astronaut, he became the first man to carry out a spacewalk without the assistance of a lifeline, relying instead on the Martin Marietta Manned Manoeuvering Unit. 'I like to think that my experience as an experimental test pilot led to my selection for this flight,' he stated later.

By the late spring of 1984 PV01 had flown at more than 51kt (96km/hr) in sideways and rearward flight. PV02 was also flying, and 12 more Apaches were in various stages of completion. In March the US Army signed contracts worth a total of $848.1 million covering 112 aircraft due to be built in FY84.

Production standard

PV03 was the first Apache built to the full production standard. Handed over to the US Army in mid-May, it was flown from Mesa to Culver City so that Hughes logistics personnel could use it to validate the maintenance manuals which Army mechanics would use.

Most of the aircraft delivered from the summer onwards were used at Mesa and Yuma for Initial Key Personnel Training (IKPT). As the name suggested, IKPT was a programme intended to train the personnel who would in turn become Army instructors. First to go were PV04 and 05: completed in July, these were the first Apaches to be handed over to the Army. They were flown to the Yuma Proving Ground for pilot weapons system and combat skills training.

Total flying hours by the prototypes and production fleet had now passed the 5,000-hour mark, while ground testing added up to another 3,000 hours. Flushed with success, Hughes announced plans to expand operations at Mesa by building a corporate headquarters and Advanced Development Center.

Training of the first batch of Army pilots ended in August. By this time PV04 and 05 had been joined by PV06 and 07, and the newly qualified personnel were soon at work training another class of Army pilots. August proved a busy month: PV02 became the first production Apache to launch a Hellfire missile, when it released a round at Yuma, Arizona, and the same month saw the retirement of an Apache veteran when, having flown for nearly 2,000 hours, AV02 came to the end of its career. Later in the year it was sent to Fort Rucker and placed on permanent display in the Army Aviation Museum. Also retired was the name Hughes Helicopters: on August 27 the organisation was formally renamed McDonnell Douglas Helicopter Company.

By the beginning of September, 20 Apaches were on the assembly line as the company worked up to a production rate of three per month, and by this time the task of validating Apache manuals had been completed, allowing PV03 to leave the Culver City hangar which had been its home for more than three months. After completing Army acceptance tests in October, it was assigned to the Army Aviation Logistics School at Fort Eustis in Virginia.

The final batch of aircraft assigned to IKTP training consisted of PV09, 11 and 12, but PV08 and 10 were not included, having been selected for another important early task. Apache was not expected to fight unassisted: as the Deployment chapter of this book explains, the Hughes aircraft was due to work in collaboration with the Army's OH-58D scout helicopter, the role of the smaller machine being to locate and designate targets for Apache. The OH-58D was an improved version of the Cayuse being developed under the Army's Advanced Helicopter Improvement Program (AHIP), and in October PV08 and PV10 were flown to the Hunter-Liggett military reservation in central California for use in operational trials of the newly developed OH-58D.

PV13 was assigned to yet another trials task. Flown to White Sands in New Mexico, it was used for electromagnetic compatability tests. Only with the completion of PV14 were Army crews finally able to climb into an Apache and fly it away for military service. Deliveries were under way at last, 12 years after the AAH specification was first drawn up.

Right: **PV02 goes through its paces at the Yuma Proving Ground in New Mexico. This base would be used to train the key personnel needed to form the heart of the planned US Army team of Apache instructors.**

Right: **PV06 joined the growing fleet of Apaches in the summer of 1984. Like PV04 and 05, it was assigned to Yuma as part of the Initial Key Personnel Training (IKPT) programme.**

Right: **The US Army had to wait a long time to get Apache, but 12 years after the AAH specification was first drawn up, aircraft from PV14 onwards were finally assigned to the military units.**

Structure

If your car starts making odd mechanical noises, you'd be advised to stop and find out what's wrong. If your Apache starts similar behaviour when the bad guys are firing at you, keep going and head for base at the first opportunity: that rugged structure was built tough enough to take punishment and still get you home. A 23mm cannon shell may have hit the rotor, oil could be pouring out of the transmission, but don't panic – Apache will make it back. Ugly-looking it may be, but it is built to take abuse and keep flying. The secret is a combination of sheer strength and skillful engineering.

From the very inception of the AAH programme the US Army has been fully aware that the AH-64A would be operating 'in harm's way': accordingly, it has specified what must surely be the most survivable helicopter ever developed. Experience in Vietnam showed how vulnerable helicopters can be to even light anti-aircraft fire, so much design effort has been put into ensuring that the AH-64 will be the A-10 of the rotary-wing world – armoured to survive hits which would down a more conventional design.

The 1973 Middle East War taught the battle-hardened Israeli Air Force to fear the deadly firepower of the Soviet ZSU-23-4 Shilka self-propelled anti-aircraft tank, but critical areas of Apache such as the main rotor blades and cockpit are designed to take a hit from a 23mm HEI (High-Explosive Incendiary) cannon shell fired by Shilka's ZSU-23 cannon without detracting from the aircraft's ability to fly.

For most practical purposes, the remainder of the aircraft is invulnerable to API (Armour-Piercing Incendiary) rounds fired from 0.5in (12.7mm) heavy machine guns. Although much smaller and less powerful than the 23mm cannon, the 0.5in machine gun is still a destructive weapon: it formed the armament of US jet fighters as late as the F-86 Sabre and provided defensive fire for early-model B-52 bombers, and in its infantry version the 'Big Fifty' was recently returned to production to meet the demand for weapons of this type.

With its insect-like profile, Apache is unlikely to win prizes for beauty, but that ugliness is strictly functional. The aircraft has a small surface area for its size, while its tadpole-like fuselage configuration requires the aircrew to expose a minimum of their mount when engaging targets. Its angular shape was partly dictated by the armour which went into its construction – valuable protection for the two-man crew.

Armour may be able to protect some of the vital areas, but the weight which it adds prevents large-scale application. Only the softest and most vulnerable items can be given such protection – in practice, the crew plus vulnerable points on the engines – while hydraulic and oil lines are protected by Kevlar composite material. The latter takes the form of a trough or gutter in which the lines are run. The rest of the aircraft and its systems are exposed to possible fire, and must be protected in some other way.

Damage tolerance

In the motion picture *Lawrence of Arabia* the hero is asked to explain his ability to extinguish lit matches with his fingertips. A soldier who tries to emulate the trick complains that it hurts, only to be told that the secret lies in 'not minding that it hurts'. The same principle applies to Apache: vital components must either be tough enough to withstand hostile fire, or be designed to operate even when damaged.

One method of making components tough (adding 'ballistic tolerance' in the words of Apache engineers) is to deliberately over-size. For example, drive shafting 3in (7.5cm) in diameter would be able to handle the normal flight loads, but by making it 7in (17.75cm) in diameter designers gave it the ability to take a 12.7mm strike and keep functioning. Around 2,500lb (1,130kg) of airframe weight is for ballistic protection, the result either of armour being fitted or of components being oversized. Despite this weight penalty, the aircraft is some 300lb (135kg) lighter than the Army-specified maximum.

Apache's fuselage is of conventional semi-monocoque construction, and is

Above: Apache may be a McDonnell Douglas Helicopter product, but most of its fuselage is built by Teledyne Ryan, then shipped to Mesa where it will be fitted out as an operational attack helicopter.

Below: Technicians install cockpit wiring and equipment. Note the large diameter of the vertical shaft which drives the main rotor – oversizing of critical components allows Apache to absorb combat damage.

Below: Technicians use a digital test meter (the pale-coloured box lying ahead of the canopy) to check out Apache's electrical wiring during the installation of avionics units including the TADS/PNVS.

manufactured from aluminium alloy by Teledyne Ryan Aeronautical, the company which supplies most of the AH-64A structure. The engine bays are widely spaced in order to prevent both T700 turboshafts being damaged by a single hit, and to reduce the risk of an explosion in one engine damaging the other, while infra-red suppressors on the engine exhausts prevent the exhaust plume from offering a good target to heat-seeking missiles such as the SA-7 Grail and SA-9 Gaskin.

Ceradyne provides lightweight boron armour which is located in the sides and floor of the cockpit, and between the front and rear seats; this can withstand strikes from 23mm HEI projectiles. Simula manufactures the armoured crew seats from Kevlar material.

Canopy transparencies are supplied by PPG Industries, while Sierracin is responsible for the transparent acrylic blast barrier between the front and rear cockpits. The windscreen is electrically heated, and flat canopy transparencies give the canopy an angular appearance but reduce glint. Conventional curved transparencies tend to act as fairly effective glint sources over a range of aspect angles, so can betray the position of an aircraft, whereas individual sections of a flat-panel canopy reflect sunlight in one direction only and are consequently less likely to draw hostile attention.

The front seat has a folding cyclic stick, a collective lever, and rudder pedals, plus a basic set of flight instruments. Visibility is good, although the

Above: Apache crew lean forward for pre-takeoff checks. Note how the armoured side of the seat shields the gunner's torso. When both crewmen sit back in their seats, protection will be more effective.

gunner cannot see very much to the rear. The pilot's seat is raised by 19in (46cm), giving him a better view. The gunner is a qualified pilot, and if a 23mm strike were to kill one crew member, the blast shield between the two cockpits should protect the other, allowing him to fly the aircraft back to base.

The tail section is of bolted pylon structure. Also built by Teledyne Ryan, it incorporates the all-moving horizontal stabilator which, driven by Simmonds actuators and Hamilton-Standard control electronics, can move from $+25°$ to $-5°$ of incidence. The $+25°$ setting is used to minimise downthrust from the tail during the hover, and is steadily reduced once airspeed rises above 30kt (55km/hr). By around 60kt (70km/hr) it is almost horizontal, and responding to collective lever position, airspeed, and pitch-attitude rate.

Cantilever wings

Apache's stub wings are cantilever mid-mounted units of low aspect ratio. Again built by Teledyne Ryan, they are removable and can be attached to the cockpit sides when the aircraft is being stored or transported. The hard points under each wing are plumbed for fuel, allowing Apache to carry four external tanks. In addition to carrying stores, the wing also acts as a vibration absorber, particularly when wing stores are fitted.

The aircraft is able to operate from forward areas, being maintained without traditional aids such as stands and work platforms. Engine cowlings can be lowered for use as work platforms, a catwalk runs along the top of the fuselage, and crew steps are provided in the vertical fin.

Two crash-resistant fuel tanks in fuselage have a total capacity of 313 Imperial gallons (376 US gallons, or 1,422 litres). Both tanks are self-sealing, and have been designed to cope with hits from the 14.5mm heavy machine gun used by the Warsaw Pact. The aircraft can be fitted with external fuel tanks in order to pro-

Impact resistance

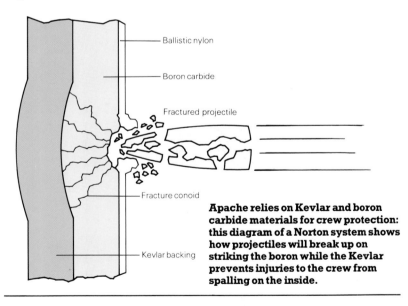

Apache relies on Kevlar and boron carbide materials for crew protection: this diagram of a Norton system shows how projectiles will break up on striking the boron while the Kevlar prevents injuries to the crew from spalling on the inside.

Cockpit protection

In flight, pilot and gunner will be protected by the armoured windscreen, as well as armour built into the seats and the sides and floor of the fuselage. Blast/fragment shielding plus a transparent blast screen divide the cockpit into two halves, so that a single hit is unlikely to harm both crewmen.

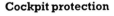 Crew compartment armour

 Blast/fragmentation shield

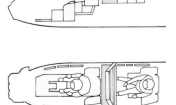

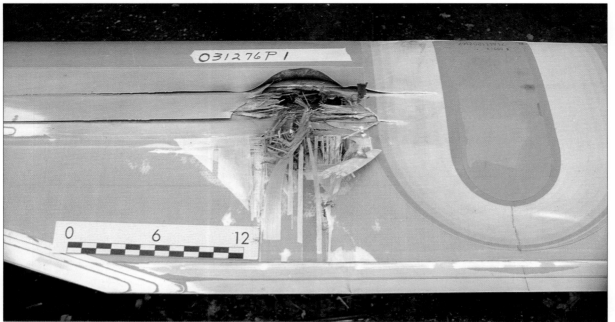

Left: After a cannon shell had burst open the titanium leading edge skin and severely damaged the glass fibre interior, this Apache main rotor blade was able to survive more than five hours of load testing.

vide the range needed for self-deployment from the continental USA to Europe or other parts of the world. These are made from Kevlar, and have a capacity of 230 US gallons; four can be carried beneath the stub wings.

Instead of the standard high-pressure inlet type of fuel system, the T700 has a suction system with a boost pump. This reduces fire hazard, as does a nitrogen inerting unit devised by Clifton Precision Instruments.

The main transmission combines the power from the two nose gearboxes, using this to drive the main rotor, the drive shaft to the tail rotor, and accessories such as electrical generators and hydraulic pumps. Power from the output shaft of the T700 engine is transferred to the rotors via a drive train consisting of the engine nose gearbox and main transmission. Designed to reduce shaft speed and increase torque, these are manufactured by Litton Precision Gear.

During the Vietnam War almost all the major components in the drive train of the UH-1/AH-1 series were changed at 1,100 flight hours, and with aircraft flying up to 100 or 130 hours per month the maintenance task was enormous. 'We were just flying the aircraft day in and day out', recalls one former AH-1 pilot, 'In every year you were changing the drive train system on every aircraft out there – it got to be awfully expensive and time-consuming'. For Apache, the Army specified 4,500 hours for component life.

Transmission hardness

The transmission has always been a vulnerable point in any helicopter, and could be damaged even by small arms fire: in Vietnam aircrew soon learned that when in combat, it paid to remain on the alert for possible transmission damage. At the first sign of rising oil temperature or unusual behaviour, pilots would break off the action, setting the aircraft down in the first safe area which could be found.

On Apache, however, the drive shafting is resistant to 0.5in rounds. The engine nose gearbox as well as the transmission and its associated gearboxes are all designed to run dry for 30mins, enough to get you home. The main gearbox can operate for up to one hour without oil, so that a hit in this vital area will not automatically result in the loss of the aircraft.

In an emergency the engine can run for up to six minutes at power settings of above 75 per cent. The nose gearbox is oil lubricated, but has demonstrated an ability to run without oil for 30 minutes. This is managed by a technique known as 'wicking', some materials used with the gearbox being designed to retain some liquid lubricant at critical bearings.

The main transmission is oil-lubricated and incorporates dual sumps separated by a baffle, so that in the event of a hit only half of the oil will be lost; should the hit damage both sumps the transmission can run dry for up to 60 minutes. Wicking is used to some degree, but most of the emergency lubrication comes from small oil-retaining reservoirs cast directly into the transmission box. The movement of air within the

Left: A main gearbox is hoisted into place for installation in Apache. Much ingenuity has been expended in the design of this critical component, which can run for half an hour after losing all its oil.

Right: The engine cowlings and stub wings are designed to act as work platforms, giving ground crew good access to the engine and transmission while eliminating the need for ladders and stands.

transmission and the rotation of the gear elements splashes this trapped oil into a mist which is sufficient to keep the unit running for the specified 60 minutes.

In 1982 Hughes started the Second Source Transmission Programme. This was intended to provide an alternative source of transmission components and gear boxes, thus keeping costs down and ensuring a smooth and regular delivery of parts to Mesa, particularly as the Apache production rate builds up to 12 aircraft a month. The first complete engine nose gearbox and main transmission to be built under this programme completed 200 hours of qualification testing in 1985, clearing these Hughes-built components for use on the production line. The work was carried out at a new $1.7 million transmission assembly and test area built at Culver City; by the autumn of that year 22 engine nose gearboxes had been built and acceptance tested, and the first five shipsets reached Mesa in October.

Static mast

Eight mechanical links hold the mast-support structure to the airframe, providing a very rigid platform for the mast, and a conical static mast surrounds the central drive shaft which powers the rotor. This configuration has several advantages. For a start, flight loads are transmitted to the airframe via the static mast rather than the drive shaft and transmission, an arrangement which helps create an agile aircraft. Secondly, the main transmission is mounted directly on the base of the mast-support structure, and can be changed in an hour without detaching the main rotor or disconnecting any of the upper flight-control systems.

Vibration levels in a helicopter cockpit tend to be high, and are fatiguing – for example, vibration can make it hard for the crew to read instruments – which can be a problem in combat. In Vietnam, it was not unusual for AH-1 crews to fly up to 10 hours per day, taking off at first light and making their last landing at dusk.

Most helicopters use vibration-damping systems of one type or another: the UH-60 Blackhawk, for example, carries vibration absorbers weighing around 300lb (135kg). These must be adjusted to tune out vibrations. When developing the Hughes 500 in the late 1960s, company designers took the bold step of eliminating vibration damping. This worked well, so the decision was taken to do the same with Apache. The two rotor systems have much in common, although the AH-64 main rotor turns at a much lower rate than that on the 500, so in theory poses a greater vibration damping problem. On the 500, any residual energy from the fast-turning blades amounts to little more than a buzz, but on the Apache it would be perceived as vibration. Absence of shock mounts or dampers helps maximise control responsiveness.

The main rotor is a four-bladed fully-articulated unit. Blades are manufactured by TRE Composite Structures Division, while the main rotor hub and flight controls are by Fenn Manufacturing. All upper flight controls are 23mm

Right: Apache's main rotor is of the fully-articulated pattern – a conservative choice. Built-in hinges allow the blades a degree of freedom to twist, rise or fall above the rotor plane, or to lag behind the rotor.

tolerant, as are the rotor blades. Each blade is 22ft (6.7m) long, and has a high-camber aerofoil section of broad chord, plus swept tips. The first design of blade to be flown was of all-metal construction. This successfully completed whirl trials on a tower test rig in February 1980. The definitive design is made of metal and composite material, and completed its whirl testing in June 1980.

On most rotor blades, around a third of the structure is the main load-bearing element, the remainder being a lightweight afterbody. About 60 per cent of the Apache blade is structure, a proportion dictated by the need to absorb 23mm hits. The remainder is made from Nomex honeycomb and afterbody. Main section of the blade consists of five stainless-steel spars lined with structural

Static mast

The conical stationary mast carries the flight loads from the rotor to the airframe, and allows the main transmission to be changed without disturbing the main rotor or the upper flight control systems.

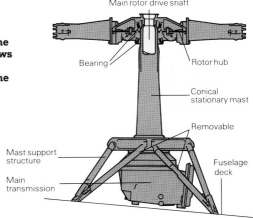

Left: Even the highest points on the aircraft can be reached without ladders. Built-in footrests allow ground crew to climb the tail fin in order to gain access to the tail rotor and its associated gearboxes.

glass fibre tubes. These effectively divide the blade into four segments. Blades have elastomeric lead/lag dampers and offset flapping hinges. They are linked to the hub by a laminated strap retention system, but can be folded or detached for transportation.

Blade protection

The sandy environment of Vietnam proved punishing on rotor blades, particularly those of the fast-turning and lighter tail rotor. In recent years blades have been treated with a paint-on erosion-preventative which can be stripped off and re-applied as necessary. A better technique is to fabricate the leading edge from a tougher material.

Current AH-1 blades incorporate a leading edge made from hard butyl rubber. This proved an effective cure for sand erosion, but problems with rain erosion have been reported. On Apache the main rotor and tail-rotor blades have titanium leading edges.

The blades are designed to last at least 4,500 flight hours and survive hits from enemy gunfire and contact with tree limbs, and during tests the main rotor blades have survived direct hits from 23mm shells. The Army requires that the rotor blade will continue to operate for 30 minutes after a single 23mm hit, but trials have demonstrated the ability of a damaged blade to run for more than five hours.

The tail rotor is mounted on the port side of the vertical tail. It is a four-bladed unit of 55/125° scissors configuration, rather than the more common right-angled cruciform pattern. The blades are manufactured by TRE Composite Structures Division.

The drive shaft to the tail rotor runs along the top of the tail boom and is covered by a simple fairing intended to keep dirt out. An intermediate gearbox at the base of the vertical fin drives a shaft which runs up towards the rotor, where a 90° gearbox transfers energy to the tail rotor.

For the first time the Army is using

McDonnell Douglas AH-64 Apache

1 Night systems sensor scanner
2 Pilot's Night Vision Sensor (PNVS) infra-red scanner
3 Electro-optical target designation and night sensor systems turret
4 Target acquisition and designation sight daylight scanner (TADS)
5 Azimuth motor housing
6 TADS/PNVS swivelling turret
7 Turret drive motor housing
8 Sensor turret mounting
9 Rear view mirror
10 Nose compartment access hatches
11 Remote terminal unit
12 Signal data converter
13 Co-pilot/gunner's yaw control rudder pedals
14 Forward radar warning antenna
15 M230A1 Chain Gun barrel
16 Fuselage sponson fairing
17 Avionics cooling air ducting
18 Boron armoured cockpit flooring
19 Co-pilot/gunner's "fold-down" control column
20 Weapons control panel
21 Instrument panel shroud
22 Windscreen wiper
23 Co-pilot/gunner's armoured windscreen
24 Head-down sighting system viewfinder
25 Pilot's armoured windscreen panel
26 Windscreen wiper
27 Co-pilot/gunner's Kevlar armoured seat
28 Safety harness
29 Side console panel
30 Engine power levers
31 Avionics equipment bays, port and starboard
32 Avionics bay access door
33 Collective pitch control lever
34 Adjustable crash-resistant seat mountings
35 Pilot's rudder pedals
36 Cockpit side window panel
37 Pilot's instrument console
38 Inter-cockpit acrylic blast shield
39 Starboard side window entry hatches
40 Rocket launcher pack
41 Starboard wing stores pylons
42 Cockpit roof glazing
43 Instrument panel shroud
44 Pilot's Kevlar armoured seat
45 Collective pitch control lever
46 Side console panel
47 Engine power levers
48 Rear cockpit floor level
49 Main undercarriage shock absorber mounting
50 Linkless ammunition feed chute
51 Forward fuel tank; total fuel capacity 375 US gal (1,419 lit)
52 Control rod linkages
53 Cockpit ventilating air louvres
54 Display adjustment panel
55 Grab handles/maintenance steps
56 Control system hydraulic actuators (three)
57 Ventilating air intake
58 UHF aerial
59 Starboard stub wing
60 Main rotor blades
61 Laminated blade-root attachment joints
62 Vibration absorbers
63 Blade pitch bearing housing
64 Air data sensor mast
65 Rotor hub unit
66 Offset flapping hinges
67 Elastomeric lead/lag dampers
68 Blade pitch control rod
69 Pitch control swashplate
70 Main rotor mast
71 Air turbine starter/auxiliary power unit (APU) input shaft
72 Rotor head control mixing linkages
73 Gearbox mounting plate
74 Transmission oil coolers, port and starboard

Apache exposes its secrets in this detailed cutaway drawing. Compared with a jet fighter the structure seems simple; much of the aircraft's massive price tag results from the complex avionics suite needed to detect and engage targets in all weathers and at night. These black boxes fill the extreme nose section and the fuselage sponson fairings.

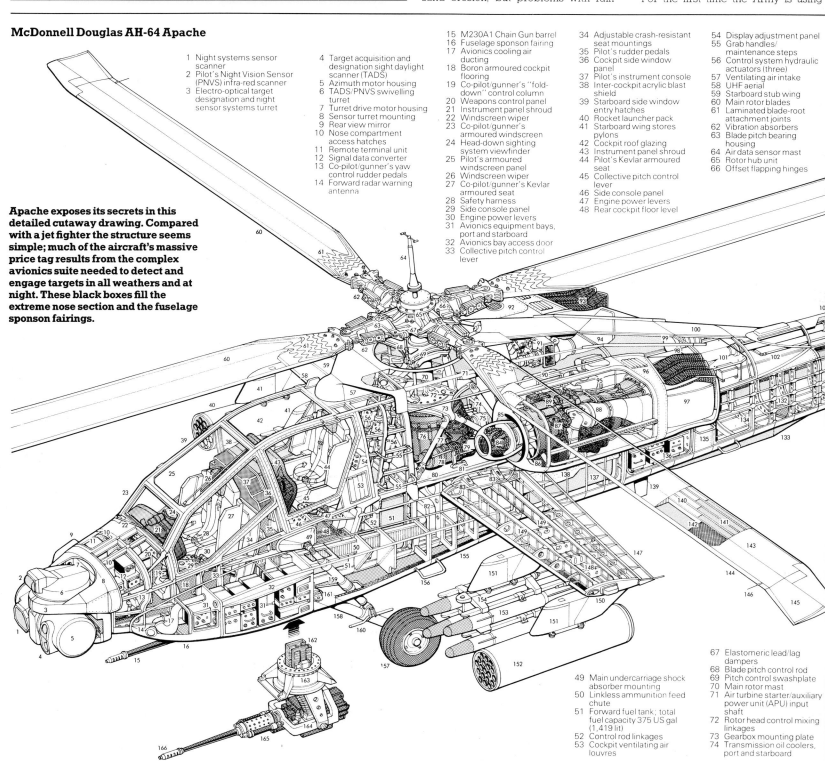

grease-lubricated gearboxes in order to give the transmission a run-dry capability. At normal running temperature the grease becomes more viscous, but is not as fluid as oil, and in the event of combat damage to either gearbox it will tend to stay in place. These boxes can take hits from 12.7mm AP rounds and still operate for the required period. Tests have shown that a damaged tail gearbox can run for up to two hours.

Main and tail rotors are de-iced by heater blankets supplied by Sierracin, while the West German company AEG-Telefunken provides the de-icing control system.

All landing gears are of the Menasco trailing-arm type, and the single mainwheels are fitted with hydraulic brakes. Although non-retractable, they are designed to fold rearward in order to reduce storage/transportation height. The tailwheel is fully-castoring and self-centring.

Ground stability

Despite being tall, with a main gear of narrow track, the undercarriage provides good ground stability. Aircraft can operate on slopes of up to 12° in the nose/tail direction and 10° sideways.

Descent rates of up to 10ft/sec (3m/sec) are acceptable in normal operations. A descent speed of 12ft/sec (3.65m/sec) will cause the undercarriage struts to blow nitrogen pressure from their cylinders, allowing them to collapse and absorb the impact energy. A higher descent rate will blow the tyres, while higher rates still will initiate structural crushing, allowing the crew to survive rough landings at up to 42ft/sec (12.8m/sec).

To help with the integration of components and subsystems within the airframe, Hughes used a skeleton version of the aircraft. Delivered to the company in August 1981, this was known as the System Development Fixture. The rig gave engineers good access to all locations within the fuselage, allowing engineers to devise the best location for equipment and cabling in the production aircraft.

The Parker Bertea dual hydraulic system operates at 3,000lb/sq in (207 bars), powering actuators ballistically tolerant to 0.5in hits. Garrett provides the aircraft's auxiliary power unit, while the two 35kVA engine-driven AC generator, control unit and transformer rectifier are by the power division of Bendix.

The integrated pressurised air subsystem supplied by Garrett AirResearch includes a shaft-driven compressor, air turbine starters, pneumatic valves, temperature control unit and environmental control unit. Cooling air passes first to the avionics, then to the cockpit. For comfort reasons the crew might prefer the opposite – the Apache canopy acts as an efficient greenhouse – but avionics units are less tolerant of overheating than humans.

Pressurised air from Apache's APU plus nitrogen from an on-board gas bottle is used to start the engines one at a time via the air-turbine starters. Once one engine is running, a cross-bleed system allows one to start the other. In buddy start mode, an air line from the pressurised air system of one Apache can be used to start the engines of another, a technique designed for combat recovery of a damaged aircraft. A ground power unit could also be used.

In the event of a crash, the Apache structure is designed to give its crew the best possible chance of survival, and a 95 per cent chance of surviving a crash landing at a sink rate of up to 42ft/sec (12.8 metres/sec). Sink rates and specification values make dry reading and the measure of protection which Apache provides can best be understood by converting these figures into more commonly used units: essentially, Apache's crew should survive an impact with the ground at a speed of close to 30mph (48km/hr). A motorist who walked away from a crash at such speeds would consider himself lucky; an Apache crew who didn't might feel aggrieved!

During a crash landing, the landing gear and lower fuselage structure will collapse and crush, absorbing the impact loads, and the designers have taken care to ensure that as the structure deforms, the belly-mounted Chain Gun avoids the crew compartment. The static mast is designed to retain the rotor during a crash and will act as a makeshift roll bar if the aircraft turns over, protecting the crew from the threat of a collapsing canopy.

Aircrew survival

The theory is excellent, the practical results equally so. The US Army has already seen crews survive uninjured the sort of accident which in the past would have involved blood wagons and body bags. Early in 1986 an Apache taking off from Fort Rucker suffered an accidental engine shut-down just after leaving the hover, and landed in trees on its left side. Acting like a giant rotary-bladed lawnmower, the main rotor carved down more than fifty trees, but when the noise of splintering wood and crushing aluminium died away, the crew were able to walk away from the disaster.

Hughes engineers had been told to build Apache tough – and they did. Accidents are unavoidable, particularly in nap-of-the-earth fight, but by the time the last Apache goes to the breakers some time in the twenty-first century many former aircrew will owe their lives to the skill of the men who designed its structure.

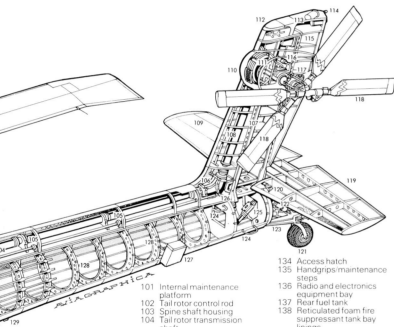

Above: The tail rotor is not a single four-bladed unit, but consists of two separate but rigidly connected two-blade assemblies whose angular position is intended to minimise the noise signature.

75 Rotor brake
76 Main gearbox
77 Gearbox mounting struts
78 Generator
79 Input shaft from port engine
80 Gearbox mounting deck
81 Tail rotor control rod linkage
82 Ammunition magazine, 1,200 rounds
83 Stub wing attachment joints
84 Engine transmission gearbox
85 Air intake
86 Engine integral oil tank
87 General Electric T700-GE-701 turboshaft
88 Intake particle separator
89 Engine accessory equipment gearbox
90 Oil cooler plenum
91 Gas turbine starter/auxiliary power unit
92 Starboard engine cowling panels/fold-down maintenance platform
93 Starboard engine exhaust ducts
94 APU exhaust
95 Pneumatic system and environmental control equipment
96 Cooling air exhaust louvres
97 Particle separator exhaust duct/mixer
98 "Black Hole" infra-red suppression engine exhaust ducts
99 Hydraulic reservoir
100 Gearbox/engine bay tail fairings
101 Internal maintenance platform
102 Tail rotor control rod
103 Spine shaft housing
104 Tail rotor transmission shaft
105 Shaft bearings and couplings
106 Bevel drive intermediate gearbox
107 Fin/rotor pylon construction
108 Tail rotor drive shafts
109 All moving tailplane
110 Tail rotor gearbox housing
111 Right-angle final drive gearbox
112 Fin tip aerial fairing
113 Rear radar warning antennas
114 Tail navigation light
115 Cambered trailing edge section (directional stability)
116 Tail rotor pitch actuator
117 Tail rotor hub mechanism
118 Asymmetric (noise attenuation) tail rotor blades
119 Tailplane construction
120 Tailplane pivot bearing
121 Castoring tailwheel
122 Tailwheel shock absorber
123 Tailwheel yoke attachment
124 Handgrips/maintenance steps
125 Tailplane control hydraulic jack
126 Fin/rotor pylon attachment joint
127 Chaff and flare dispenser
128 Tailboom ring frames
129 Ventral radar warning antenna
130 Tailcone frame and stringer construction
131 UHF antenna
132 ADF loop antenna
133 ADF sense antenna
134 Access hatch
135 Handgrips/maintenance steps
136 Radio and electronics equipment bay
137 Rear fuel tank
138 Reticulated foam fire suppressant tank bay linings
139 VHF antenna
140 Main rotor blade stainless steel spars (five)
141 Glass-fibre spar linings
142 Honeycomb trailing edge panel
143 Glass-fibre blade skins
144 Trailing edge fixed tab
145 Swept blade tip fairing
146 Static discharger
147 Stub wing trailing-edge flap
148 Stub wing rib construction
149 Twin spar booms
150 Port navigation and strobe lights
151 Port wing stores pylons
152 Hydra 70 rocket pack: 19 2.75-in (70mm) FFAR rockets
153 Rockwell Hellfire AGM-114A anti-tank missiles
154 Missile launch rails
155 Fuselage sponson aft fairing
156 Boarding step
157 Port mainwheel
158 Main undercarriage leg strut
159 Shock absorber strut
160 Boarding steps
161 Main undercarriage leg pivot fixing
162 Ammunition feed and cartridge case return chutes
163 Gun swivelling mounting
164 Azimuth control mounting frame
165 Hughes M230A1 Chain Gun 30mm automatic cannon
166 Blast suppression cannon muzzle

Below: Adjustments are made to Apache's tail wheel. Located at the extreme end of the tail boom, this gives the aircraft a much longer wheelbase than that of the rival Bell YAH-63 design.

Powerplant

The Army wanted an engine which would be cheaper to run and simpler to maintain than any other in its performance class, and so the General Electric T700 was born: created for the UTTAS transport helicopter programme, it was the right engine at the right time for Apache. Tests have shown that it can cope with hazards such as rain, snow, ice, salt spray and altitude, delivering the power needed to allow Apache to carry a heavy warload under all likely tactical conditions. And the T700 is so reliable that its service record persuaded the US Army to throw away the warranty after only three years of production.

In 1967 the US Army initiated a competition for the development of an advanced technology demonstrator helicopter engine; intended to apply technology developed for large engines to a smaller unit, the programme lasted four years. Specific goals which the Army hoped would be achieved by this infusion of technology included reduced engine weight and lower maintenance demands. The service also hoped to obtain an engine which would be less fuel-thirsty at reduced power settings — a powerplant suitable for use in the planned UTTAS helicopter.

Development of what was then known as the GE-12 started in mid-1967, and by the middle of 1970 the design had passed its 10-hour qualification test. During the trials engines ran for more than 300 hours, demonstrating that the new GE offering met all design objectives; in December 1971 the Army announced that the General Electric design had been selected as the UTTAS powerplant, and in March 1972 General Electric received a contract for full-scale development.

Soon afterward the engine's first application was born: in August 1972 the US Army chose Sikorsky and Boeing Vertol to develop rival Utility Tactical Transport Aircraft System (UTTAS) helicopters. The resulting YUH-60 and YUH-61 were both powered by T700 engines.

The first T700 began running in February 1973, a few months before the contract to build the YAH-63 and YAH-64 prototypes was awarded to Bell and Hughes, and in February 1974 GE delivered the first units for ground testing. Preliminary flight rating test (PFRT) work was completed in July 1974, allowing delivery of flight-rated engines to begin in the following month, and these were quickly installed in the rival UTTAS prototypes. Sikorsky's YUH-60A flew for the first time on October 17, 1974, starting the T700 off on its operational career. The rival Boeing Vertol YUH-61A started flight tests just over a month later on November 29. In September and November of the following year, the engine flew in the Hughes and Bell AAH prototypes. Qualification testing was also under way, and the 150-hour Model Qualification Test was completed in February 1976.

T700 first order

On December 23, 1976, Sikorsky was declared winner of the UTTAS competition, and GE received the first T700 production order. This covered manufacture of the T700-GE-700 engines with a continuous rating of 1,258shp (938kW) and an intermediate rating of 1,560shp (1,163kW). Deliveries started in March 1978. The first production example was delivered 11 years after the awarding of the initial US Army contract for an advanced technology demonstrator helicopter engine. It also marked a milestone for GE — a quarter of a century had passed since the company started work on the first US turbine helicopter engine.

The US Navy's version of UTTAS was the Sikorsky SH-60B Seahawk. For this application the Navy wanted around 10 per cent more power, so GE was given a $547,000 contract to create the T700-GE-401. Additional power was obtained by increasing the engine operating temperature, increasing the efficiency of the centrifugal compressor and by reducing air leakage. Continuous rating is 1,437shp (1,071kW), while intermediate

Above: Low fuel consumption, light weight and reduced maintenance demands were key requirements for General Electric's T700 turboshaft, making this engine the obvious powerplant for Apache.

Below: This may be a posed publicity photo, but it illustrates one important feature of the T700 — the minimal tool kit needed for front-line maintenance. Most LRUs can be changed using 10 common tools.

Above: Early-model T700 turboshafts first flew in 1974 on the rival Boeing Vertol and Hughes Utility Tactical Transport Aircraft System (UTTAS) prototypes.

rating is 1,693shp (1,261kW) and contingency rating is 1,729shp (1,289kW). Corrosion resistance has also been improved to suit naval operations: by the summer of 1980 the -401 had completed 800 hours of salt water testing.

GE engineers used the -700 as the starting point when designing the AH-64 powerplant, and the resulting T700-GE-701 was essentially an Army version of the Navy's -401. Its external dimensions and weight are almost identical with those of the earlier models, the main engineering change being the use of Inconel 718 rather than steel for the compressor diffuser, and commonality between the Apache and Blackhawk powerplants is greater than 95 per cent. Length of the -701 is 46in (116.8cm), and the maximum diameter is 25in (63.5cm). Weight is 437lb (198kg), identical to that of the -700 and 3lb (1.4kg) heavier than the Navy's -401 version.

Like most modern aero engines, the T700 can be broken down into self-contained modules which are functionally and mechanically interchangeable between individual engines. The T700 has four such modules, breaking down into a hot section, cold section, power turbine and controls/accessories.

The annular intake incorporates an anti-iced integral inlet particle separator. Although the latter has no moving parts, it can remove up to 95 per cent of the dust, sand and small foreign bodies in the incoming air. A separator blower driven from the accessory gearbox discharges this material over the side of the engine.

The compressor is of combined axial/centrifugal configuration, and has a compression ratio of approximately 15:1. Airflow is around 10lb (4.5kg) per second at 44,720rpm. One unique feature of the engine is that this section is made of only 11 major parts. Each of the five axial stages mounted on the engine's single shaft takes the form of a one-piece unit which combines disk and blades in a single component. Termed 'blisks' by GE, these are manufactured from AM355 steel, a material highly resistant to corrosion. The inlet guide vanes and the stator blades of the first two stages are variable, while the tips of the vanes on the centrifugal impeller are backswept and the compressor casing splits axially to allow maintenance.

Air then passes to an annular combustor of compact configuration. Fuel is injected centrally, an arrangement which gives good tolerance of fuel contamination and produces minimum smoke, while offering a uniform temperature profile at the turbine. The flame tube is of ring configuration, and machined from Hastelloy X – a nickel/chromium/molybdenum/iron alloy which combines good strength at high temperatures with excellent resistance to oxidisation and corrosion. A dual high-energy ignition system draws power from separate winding on the engine-mounted alternator.

The engine has two turbines. The high-pressure (HP) unit used to drive the gas generator section of the engine is of two-stage configuration, and operates at gas temperatures exceeding 2,010°F (1,100°C). The first-stage nozzle is investment-cast in X40, while the second stage nozzle is investment-cast in two-vane segments using Rene 90. Blades, cooling plates and discs are clamped by five tiebolts, while five larger bolts attach the entire assembly to the shaft which drives the compressor section.

At intermediate power setting, gas leaves the HP turbine at a temperature of around 1,520°F (827°C), then enters the two-stage free power turbine used to extract power for the drive shaft. Designed to give high efficiency at lower throttle settings such as 30 and 60 per cent of military power, it has tip-shrouded blades and segmented nozzles. The nozzle guide vanes are made from Rene 77 nickel-based alloy, the rotor discs from Inconel 718. The turbine blades are manufactured from Rene 120 alloy by precision casting, and given a nickel-alumide coating. The efflux is then passed to an annular one-piece exhaust. Lubrication and electrical systems are self-contained, reducing the engine's dependence on airframe services.

Power ratings

Maximum power rating (intermediate power) of the T700 is normally 1,698shp (1,266kW), a level which can be maintained for up to 30 minutes, while maximum continuous rating is 1,510shp (1,126kW). If one engine fails, the other can produce 1,723shp (1,285kW) for two and a half minutes, a contingency rating which should allow the aircrew to break off their mission and carry out an emergency landing.

The above figures refer to sea level static performance at an temperature of 59°F (15°C). In the demanding 4,000ft/95°F (1,220m/35°C) conditions specified by the Army, the T700-701 has an intermediate rating of 1,310shp (970kW), a maximum continuous rating of 1,069shp (7097kW) and a contingency rating of 1,443shp (1,076kW).

The engine operates with no visible smoke and at low noise levels. At maximum take-off power the HP shaft rotates at 44,720rpm, while the front-drive power output shaft rotates at speeds of between 17,000 and 21,000rpm. Matching of engine speed and torque is managed by electrical and hydromechanical control.

Specific fuel consumption (SFC) of the T700-700 is between 0.464 and 0.497lb/hr/shp (0.282-0.302kg/kW-hour), depending on operating conditions. This compares with a typical value of 0.60lb/hr/shp (0.36kg/kW-hour) for the T53-L-703 1,485shp (1,106kW) engine used on the Bell AH-1S Hueycobra, for example. SFC of the T700 is 25-30 per cent better than that of comparable engines, says GE, and the resulting fuel savings can be as much as 100,000 gallons per engine during 5,000 hours of running.

By the time all the T700s in the field had completed a combined quarter of a million hours of running, total line maintenance expended in their support was less than one man-year. For every maintenance man-hour, the engine had run for around 30 hours. By the autumn of 1984 the engine had passed the 500,000-hour mark, and by mid-1986 it had accumulated more than a million flight hours and high-time engines had comfortably exceeded 5,000 flight hours.

US Army reliability and maintainability goals have been consistently met or exceeded. Field maintenance is 75 per cent less than that required by other

Below: US Army experience has shown that engine accessories on the T700 turboshaft can be removed and replaced in between two and 15 minutes, well under the specified goal of between four and 25 minutes.

Below: The engine used on the Apache is the T700-GE-701, an uprated model developing around nine per cent more power than the T700-GE-700 used in the Sikorsky UH-60 Blackhawk.

Left: Apaches fly in formation over desert terrain. The twin T700 powerplants give the aircraft performance to spare even in the worst-case hot and high conditions which the RDF could expect to meet.

engines, says GE, while field operating experience has shown that unscheduled maintenance is running at around one third the amount required by earlier engines of comparable power.

All the modules into which the T700 breaks down are completely unitized including bearings, and the changing of any individual module involves no critical dimensional or calibration checks. Tests have shown that a two-man Aviation Intermediate Maintenance (AVIM) team can remove and replace a hot section in just under an hour, a cold section in less than 90 minutes and a power turbine in little more than half an hour; replacement of the controls and accessories module takes just over 20 minutes. All this can be done using a field maintenance kit consisting of only ten items (including four spanners), plus the appropriate handling aids. Changing of the hot section requires the tool kit plus a universal sling and an adaptor which allows the engine to be mounted on existing maintenance stands.

Design for maintenance

Experience has shown that most maintenance work on any engine is performed on external controls and accessory components, so much attention was paid during the design process to making these readily accessible, and ensuring that servicing tasks are as straightforward and foolproof as possible. Engine accessories are grouped at the top of the engine, beside the engine control system, and like all lines and connectors they are positioned so as to avoid damage when the engine is being handled. Oil level sight glasses are provided on both sides of the cold section, while all fittings and clamps are designed for ease of use. Fluid fittings do not require ground crew to use torque wrenches, for example.

Aligning features have been provided to external accessories, and a single man can change many field-serviceable parts in less than 10 minutes: typical times include fuel filter or separator blower 2mins, igniters 4mins, electrical control or torque sensor 5mins, ignition unit 6mins and hydromechanical unit 8mins. Comparison with what GE coyly refers to as an 'earlier engine' is revealing. Time to change an ignition unit is six man-minutes instead of 60, while changing a fuel control takes eight man-minutes instead of 115.

Given an engine with such virtues, other applications have been quick to follow. In 1980 Bell flight-tested a US Marine Corps AH-1T HueyCobra whose normal Avco Lycoming T400-WV-402 engine had been replaced by a pair of T700-700 engines, raising the installed power from 1,970shp (1,469kW) to around 3,200shp (2,390kW). This looked promising, so Bell proposed a definitive upgrade based on the T700-401. A demonstration aircraft flew on November 16, 1983, and started trials with the USMC in the following month. The Marines were impressed, deciding to upgrade their fleet of around 40 aircraft to this standard and to purchase 44 new-build aircraft for delivery from March 1986 onward – good news for GE, but not

Left: Apache's engines are housed in widely spaced cowlings on the outside of the fuselage. A single strike is unlikely to damage both, while the fuselage protects one engine from an explosion in the other.

so good for Hughes; the latter had hoped that the Marines might buy a proposed version of the AH-64 equipped with folding blades for shipboard use.

Another application for the T700 is the US Navy's Kaman SH-2F Seasprite LAMPS (Light Airborne Multi-Purpose System) helicopter. The Seasprite is currently powered by a pair of 1,350shp (1,007kW) General Electric T58-GE-8F engines, but a YSH-2G prototype fitted with two T700-401 powerplants started flight trials in late 1984. As part of a block upgrade scheme involving improvements to avionics, some aircraft will also be retrofitted with the newer engine.

Having been adopted as the powerplant of the Apache and three other types of front-line US aircraft – the Sikorsky UH-60A Black Hawk, Sikorsky SH-60B Seahawk, and Bell AH-1T and -1W HueyCobra – the T700 is the dominant medium-powered helicopter engine of its generation. The production line at General Electric's Aircraft Engine Group at West Lynn, Massachusetts, had built more than 3,000 T700 engines by the summer of 1986; production was running at around 1,000 engines per year, and was expected to continue into the early 1990s.

In 1983 the Pentagon decided to award a multi-year contract for the T700. In October of that year GE received a contract worth $738 million – the largest ever received by the company for military engines. This covered the manufacture of 1,554 T700 engines to be supplied in FY83, FY84, and FY85 for Army, Navy, and Air Force use. By buying in bulk, the US Services stood to make substantial savings: according to GE, this massive contract cut the Pentagon's T700 engine bill by more than 10 per cent. Unit cost is probably around $500,000.

Given the military success of the T700 it was hardly surprising that GE decided to promote the engine for commercial applications. First to be announced was the CT7-1, which was launched in September 1976 and certified the following June, but the sole application of this 1,560shp (1,160kW) unit was the Sikorsky S-78, a proposed commercial derivative of the UH-60A which would have carried 20 or 29 passengers.

The CT7-1 was followed by the uprated 1,620shp (1,209kW) CT7-2A which powers civil versions of the Bell 214ST, and the 1,725shp (1,287kW) CT7-2B used in the Westland 30 Series 200 and 300. A civil derivative of the UH-60 series finally took shape in the form of the S-70C, which flew for the first time in June 1984 and is powered by a pair of 1,625shp (1,212kW) CT7-2C engines. Another application is the Bell 214ST: originally developed for Iran back in the days of the Shah, it has enjoyed significant success as a civil transport and is powered by two CT7-2A engines.

Given its widespread use by the US Services, the T700 was clearly a prime candidate for foreign programmes. In July 1980 GE signed an agreement covering the licence manufacture of the T700 by the Aero Division of Alfa Romeo at Pomigliano d'Arco, Italy. Under the agreement, the Italian company would manufacture the T700/CT-7 for use in Italian and European helicopter programmes.

EH-101 application

One obvious application was the planned European Helicopter Industries EH-101 medium-weight helicopter being developed by Agusta and Westland. The CT7-2A was selected as the powerplant for prototype aircraft, the definitive engine being the Rolls-Royce/Turbomeca RTM322. Some observers predict that production aircraft may use both powerplants, with Italy's planned fleet of 38 being T700-powered and the UK opting to fit the RTM322 on its 50 aircraft. First flight of the EH-101 is due to take place early in 1987.

The international collaboration deal also covers the development of uprated models of the T700. In January 1985 General Electric announced that its Business Engines Group would collaborate with Alfa Romeo and Fiat Aviazione to develop a new derivative of the T700/CT7. The new variant would be a suitable powerplant for medium-weight helicopters such as the civil version of the EH101.

By adding a speed-reduction gearbox driving an advanced technology 10.5ft (3.2m) diameter propeller, GE has been able to offer a turboprop version of the T700. This project was launched in July 1979 after GE studies had shown that the helicopter engine could be adapted with minimal changes to create a new turboprop engine of around 1,500shp (1,120kW) rating. An engine of this size would be suitable for use in the new generation of 30–40-passenger commuter aircraft expected to enter service in the late 1980s, and would benefit from more than 15 years of military-funded development and flight testing of compressors, combustors, and turbine components.

First turboprop version was the short-lived CT7-3 rated at approximately 1,500shp (1,122kW) but work on this version soon ceased in favour of the more powerful CT7-5, which in turn was developed into the 1,600shp (1,192kW) CT7-5A1. The company's boldness paid off: in June 1980 the -5A1 was chosen as the powerplant of the Saab Scania/Fairchild SF 340A, and in the following year the uprated 1,700shp (1,267kW) CT7-7A was adopted for the CASA/Nurtanio CN-235.

The CT7 flew for the first time in September 1982 and was certified by the FAA on August 29, 1983, entering airline service on June 15, 1984, with the SF 340As of the Swiss regional airline Crossair. Soon after their entry into service, Crossair's new fleet experienced three unrelated CT7 incidents, but GE quickly devised modifications to correct the problem. By the summer of 1985 only one shutdown of a modified engine had been reported.

More powerful CT7 turboprops are already being planned. The 2,000shp (1,490kW) CT7-9 tested in 1986 is reported to offer a 3-4 per cent improvement in SFC. An even more powerful CT7-11 in the 2,200-2,400shp (1,640-1,790kW) performance class is proposed for service in the early 1990s.

Above: Hot exhausts and engine efflux can be a prominent target for infra-red guided missiles, particularly shoulder-fired SAMs. Apache's exhausts are fitted with Black Hole suppressors.

Below: The addition of a speed-reduction gearbox to the basic T700 has resulted in the CT-7 turboprop engine used in small civil airliners such as the CASA Nurtanio 235 and the Saab-Fairchild 340.

Avionics

"You've heard of the expression, 'If looks could kill'", Sean Connery is told in the science fiction movie *Zardoz* – "Well, here they can." Apache crews may not have the ability to kill at a glance, but the helmet sights they wear can use that glance to direct the fire-control system packed into the AH-64's bug-eye nose, and once a target is found a missile or projectile which kills could be on its way within moments. That fire-control system accounts for a significant share of Apache's high cost, but plays an important role in giving the helicopter its high combat effectiveness.

No single component of the Apache avionics suite does more to make the aircraft a hard-hitting tank killer than the Martin Marietta TADS/PNVS (Target Acquisition Designation Sight & Pilot Night-Vision Sensor). Night-vision systems, laser rangefinders and laser designators have been in military service for some time, but Apache is the first helicopter to combine them into a single integrated system. The result costs around $890,000 a copy but creates a fire-control system of unparalleled effectiveness.

The key word is integrated. To have simply bolted off-the-shelf systems into Apache would have created what computer enthusiasts would call a 'kluge' – a collection of ill-matched systems which would impose a heavy workload on the user, and which might at times simply not be used due to lack of time.

A simple example of a kluge in action was provided several weeks ago when the author contacted Martin Marietta to request information on the TADS/PNVS. The request was written using a word-processing program running on an Apple II microcomputer, checked for spelling errors by another program, then converted by a third program into a form acceptable for transmission to the USA in the form of a telex. A more modern integrated program could have combined word-processing, spelling correction and telex facilities, allowing the text to be checked then transmitted at the press of a few keys.

Above: Improved lifting performance and the ability to withstand battle damage may be useful in combat, but the electronics units crammed into the nose and equipment sponsons are Apache's most valuable feature.

Left: TADS/PNVS on test. The mirror-like window over the TADS FLIR reflects an external view which includes a tank target. The latter can be seen at a larger scale in the thermal image being screened.

Below: The PNVS sensor mounted above the aircraft nose and the larger TADS sensors mounted below are independently steerable. Note how the PNVS sensor has turned to match the direction of the pilot's gaze.

Kluges in combat

In combat, kluges go unused. A good example occurred during the Vietnam War, when US Navy A-6 Intruder bombers carried two radars — the APA-92 search radar and the APQ-88 or -112 track radar. With the North Vietnamese air defences filling the sky with anti-aircraft fire and SAMs, crews were unwilling to attempt the task of operating two sets, so they often ignored their aircraft's track radar, relying on the APQ-92.

On Apache, great care has been taken to integrate the attack systems, making them easy to use. All sensors use a common boresight, so that they can swiftly be pointed in the same direction and moved in unison, while the interface between man and machine has been made as simple as possible.

Following the competitive fly-off against a rival Northrop equipment, Martin Marietta was given a contract in April 1980 to complete development of the system. This $45.5 million engineering development contract was followed by $94.6 million in FY82 plus $37.4 million for long-lead production hardware. A DSARC III (Defense Systems Acquisition Review Council III) meeting held in February 1982 approved TADS/PNVS for production, and first deliveries were made in November of the following year.

Sensors for the TADS/PNVS are mounted in the Apache's nose, the location used to house the missile sighting system on the earlier AH-1. This is one of the few features of the aircraft which seem unduly conservative. European designers tend to prefer mast-mounting – an alternative scheme which provides significant tactical advantages such as the ability to observe targets while most of the aircraft remains behind cover. The US Army subsequently adopted a mast sight for the OH-58D Combat Scout, the helicopter which forms Apache's teammate in action.

Windows for the TADS and PNVS are mounted in steerable turret assemblies,

Below: Mounted in the centre of the gunner's instrument panel are the displays for the TADS system – the single-eyepiece Optical Relay Tube and the green-phosphor CRT screen of the Multipurpose Sight System.

TADS/PNVS target acquisition

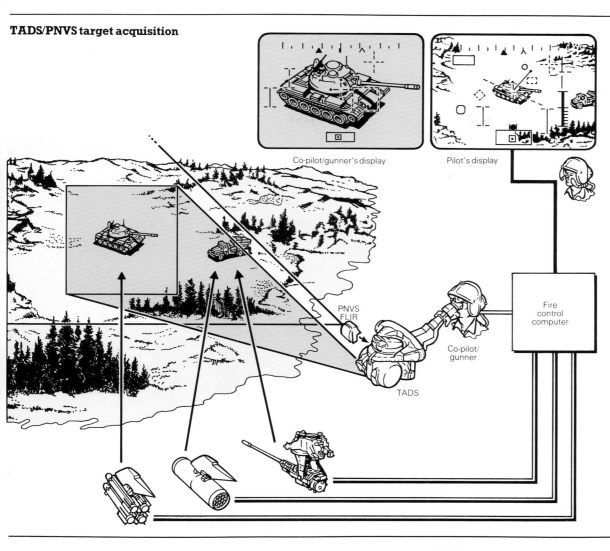

Left: PNVS gives the pilot the wide-angle thermal view which he needs in order to fly and manoeuvre by night, and to locate targets such as this tank and rocket launcher. The gunner uses the higher magnification of TADS to identify the tank target, then designate it for attack by the Hellfire missile system. Rockets or the gun can deal with the softer target.

and Display Sight System). The most obvious difference between this and most helmets is the presence of an electro-optical system mounted on the right-hand side. This protrudes upwards and across the wearer's face, ending in a small combiner glass located directly in front of the right eye. This acts like a HUD, presenting the wearer with video imagery and superimposed symbology detailing data such as airspeed, altitude and aircraft heading, enabling the pilot to fly and fight without looking down into the cockpit.

PNVS imagery

The system has no magnification, so the thermal image offered to the pilot's eye is identical in scale and directly superimposed over the external view. There is no apparent time delay between the direct view and the thermal image. Since PNVS has only a single optical channel, the thermal image is only two-dimensional, but this has not proved a problem – crews soon learn to interpret the FLIR imagery. If the pilot looks down at his cockpit instruments or displays, he faces the problem that these are at short range while the IHADSS symbology is focussed at infinity. Again, experience soon teaches him to cope with the novel sensation caused by the difference in range. PNVS can also be used by day, IHADSS symbology providing flight information and target cueing.

PNVS is slaved to the IHADSS helmet, so that movements of the pilot's head will result in a similar movement of the PNVS sensors, ensuring that the EO sensors will 'see' the same scene that the pilot does. Maximum slewing rate is 120°/sec in azimuth, and 93°/sec in elevation. Both crew members have IHADSS helmets, a

increasing the aircraft's resemblance to a giant insect and providing an unobstructed view over a large forward sector including most of the forward hemisphere and areas off the beam of the helicopter.

Sensor housings

Three housings protect the electro-optical sensors. The PNVS FLIR is contained in a single unit mounted above the nose and able to move from side to side. This may be stowed when not in use. TADS equipment is in a two-sided turret which can move in elevation and azimuth. The port-side part of the TADS assembly, often referred to as the day side, contains the TV and direct-view optics, the laser designator/rangefinder and the laser spot tracker, all mounted behind a vertical two-facet optical window. The starboard-mounted night side consists of a FLIR-mounted behind a large circular window. Based on the same IR modules as the PNVS unit, it offers three selectable magnifications.

Four TADS electronics units are mounted in the aircraft avionics bay along with a single unit for the PNVS, while the displays and control panels for the TADS and PNVS are located in the gunner/co-pilot's and pilot's cockpit respectively. Both gunner and pilot can obtain video data from either system.

TADS may lead PNVS in the equipment's designation, but PNVS is the first to be used on a mission. It is a wide field-of-view FLIR which provides the pilot with a real-time representation of the panoramic scene before him, allowing him to fly at night. The nose-mounted sensor head has a 40°×30° field of view, and can be moved horizontally to cover angles of up to 90° either side of the aircraft centreline, and vertically from +20° to −45° in elevation.

FLIR imagery can be observed using a CRT display known as the Video Display Unit, but will normally be presented to the pilot by means of the Honeywell IHADSS (Integrated Helmet

Below: The wide range of angles through which TADS and PNVS may be steered makes aircraft attitude and sensor angle look near independent of one another. The result is a high degree of tactical freedom in combat.

TADS/PNVS fields of view

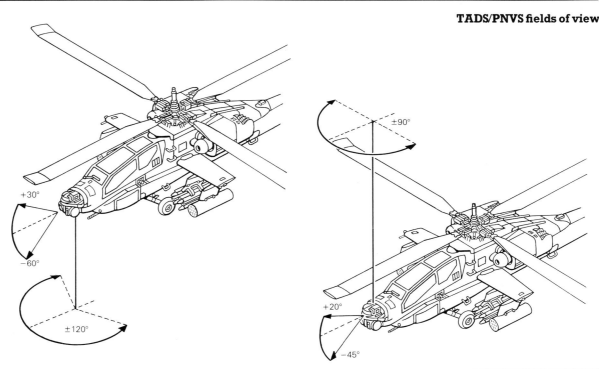

feature which allows one to 'show' a video image to the other, or to cue his partner's line of sight to an area or object of interest. Pointing accuracy is 5–10 milliradians.

Once the pilot – helped by PNVS if necessary – has positioned Apache within range of suitable targets, the gunner uses the TADS (Target Acquisition and Designation System) to acquire a target and to provide fire-control data for the Hellfire missile, 2.75in rockets or 30mm cannon. Compared with earlier optical aiming systems such as the M65 fitted to the AH-1, TADS extends the range and accuracy with which hostile tanks and AFVs can be detected, acquired and targeted.

TADS targeting

Target imagery can be obtained from any one of three turret-mounted sensors. These can be steered up to 120° on either side of the aircraft centreline, and from +30° to −60° in elevation. TADS daylight or night/bad weather sensors can be used singly, or in combinations dictated by weather or tactical conditions.

Widest field of view available from the daylight sensors is obtained by using the direct-view optical (DVO) system on its low-power (×3.5 magnification) setting. This gives a field of view 18° wide, falling to 4° if ×16 magnification is selected. The line of sight from the turret-mounted optics to the gunner is provided by the Optical Relay Tube (ORT) which terminates in an eyepiece mounted in front of him. This is designed to collapse under high g loads, a feature intended to protect the gunner's face in the event of a rough landing.

An alternative daylight sensor which trades a narrower field of view for a useful degree of smoke and haze penetration is the TADS near-infrared TV camera. This offers wide (4°) and narrow (0.9°) fields of view on the gunner's head-down Multipurpose Sight System.

For night and bad-weather attacks the gunner uses a 9in (23cm) FLIR, the largest-aperture unit of its type in tactical military service. This has three fields of view, 50°, 10.1° and 3.1° wide respectively. Water vapour in the atmosphere absorbs IR energy, so that bad weather can reduce FLIR performance, and to help cope with this degradation image intensifiers are provided in the FLIR

Below: The "day" side of TADS has two optical ports. The upper is shared by the TV system and laser, while the lower leads via a long optical path to the cockpit-mounted eyepiece and the CRT display.

Left: The optics of the IHADSS helmet sight project data and imagery onto a tiny combiner glass in front of the pilot's right eye. (The canopy stiffener seen here is not fitted to production aircraft).

system. Tests have confirmed that the unit can continue to operate at reduced performance levels in rain or snow. Stand-off attack range is reduced, but that factor works for both sides, reducing Apache's risk of detection or engagement.

Once the gunner has acquired his target using TADS, it may be tracked manually or automatically. Experience has shown that in most cases auto-tracking is more accurate than manual tracking, but can be confused if the target is momentarily obscured – by passing behind a tree, for instance. Some gunners prefer to use manual tracking during a missile engagement.

With TADS following the target in either automatic or manual mode, the International Laser Systems laser designator/rangefinder mounted in the same housing and sharing the same optics can be used to check target range, then to illuminate the target. Hellfire missiles can then be fired, or Apache can act as target designator for other laser-homing weapons such as the Copperhead smart artillery shell launched from 155mm howitzers. Another feature of the TADS system is a laser spot tracker. This can be used to locate targets marked by ground-based lasers, or lasers in other aircraft or RPVs.

Development of TADS posed significant problems with stabilisation of the sensors. These would have to keep the optics aligned even at high magnifications, coping with aircraft movement, engine and rotor vibration and the effects of gun firing. The solution adopted involves two sets of gimbals – a coarse outer gimbal works in conjunction with a finer inner gimbal coupled to rate-integrated gyros. The result is a system so stable that it can hold its boresight to within a few feet at a distance of eight miles.

Service experience has shown that Apache crews do not always use their TADS/PNVS in the rigid manner described. 'Pick-and-mix' operating methods, in which both crew members may use both systems, switching between optical, TV and IR TADS tracking as required, give a better end result and increase the chances of a successful mission. Switching between DVO or TV and IR images may give useful clues – an abandoned tank will be thermally cooler than one waiting with engine and systems running, for example.

TADS may have been installed as a target-acquisition and weapon aiming device for the gunner's use, but it may also be used in wide-angle mode to supplement PNVS during night operations, giving the aircraft a second pair of thermally-equipped eyes as it moves over the darkened battlefield. It can also be used by the pilot, either as a back-up should his PNVS fail or as a method of taking a close look at a suspect area.

Although custom-designed for Apache, TADS/PNVS could see service in a rival aircraft – the PAH-2 version of the proposed Franco/German attack/scout helicopter. In the summer of 1986 the US Defense Department formally notified Congress that it intended to offer West Germany the chance to purchase 300 mission-equipment packages for the PAH-2 helicopter. These would consist of the TADS/PNVS plus a video thermal tracker. Estimated unit cost of the proposed system is $1.24 million.

Pilot displays consist of the integrated

TADS/PNVS optical paths

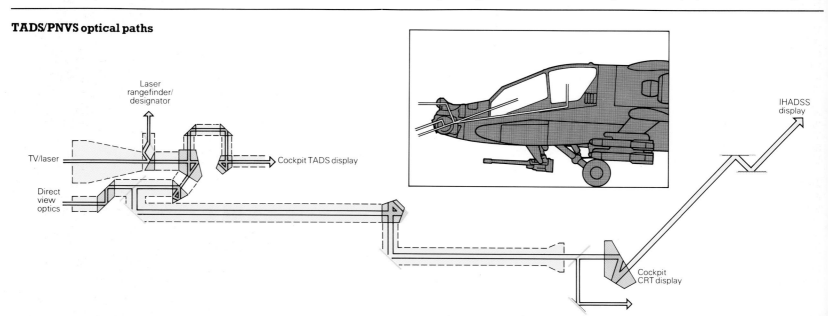

Avionics

Martin Marietta AAQ-11 Mk III PNVS turret

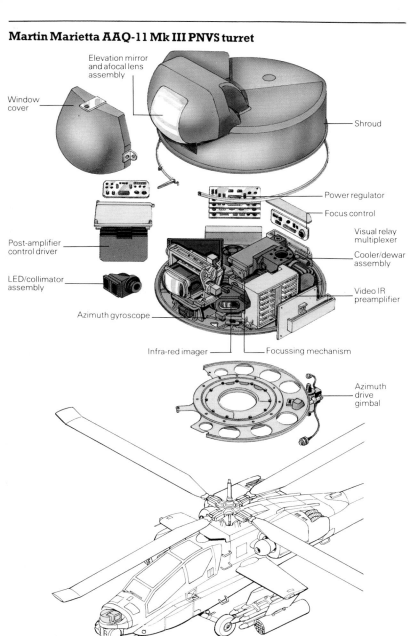

Above: Mounted at the top of the Apache's nose assembly, the AAQ-11 PNVS turret houses the pilot's wide-angle FLIR. The Mk II version has no elevation gimbal or torquer, while the Mk I also dispenses with the azimuth gimbal. The infra-red night vision system is not only vital for night flying – it can also be an invaluable aid for the pilot in the normal battlefield conditions of smoke, sand and dust clouds.

Above: As the pilot moves his head, the IHADSS helmet sends steering commands to the PNVS, and superimposes a "patch" of thermal imagery on the view of the external scene he is shown.

Below: Interconnections between the TADS/PNVS, IHADSS, SSUs (Sensor Surveying Units), DAP (Display Adjust Panel), other avionics and the cockpit displays are shown in this system block diagram.

Honeywell IHADSS components

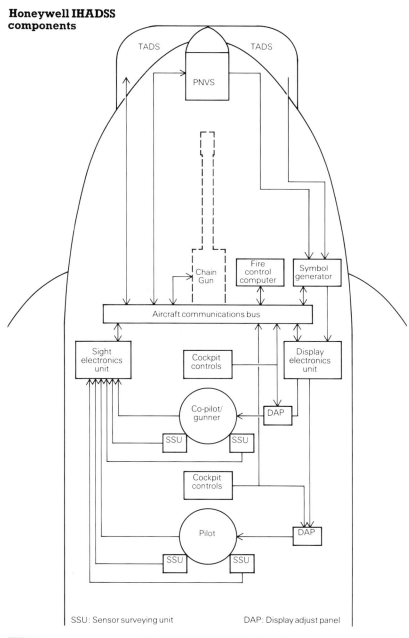

helmet display sight system helmet display unit (IHADSS HDU) and a panel-mounted video display unit. Gunner displays include the IHADSS HDU, a target acquisition and designation sight head-down display (TADS HDD) and a TADS heads-out display.

An Astronautics Corp head-down video display unit provides flight, navigation and weapons data on a 5.5in × 5.5in (14cm × 14cm) high-brightness CRT. Other displays include a horizontal situation indicator, a radio magnetic indicator and a remote altitude indicator.

Teledyne Industries supplies Apache's fire-control computer, while the external stores controller is from Textron. The Aerial Rocket Control System is supplied by the Flight Systems Division of Bendix.

Automatic stabilisation

Apache is fitted with a Sperry Flight Systems dual Digital Automatic Stabilisation Equipment (DASE). This takes information via the main databus from different parts of the aircraft. Rate, velocity and heading data comes from the inertial-quality Litton Heading and Attitude Reference System (HARS), while air data, low-airspeed and wind direction information comes from the pacer system on the rotor mast, a low-airspeed data system which indicates the azimuth and velocity of the air, plus temperature and pressure data. The resulting information is distributed via the aircraft's databus system to the various ordnance systems which require it for fire control.

Having analysed these inputs, DASE matches them to the pilot's demands and to the corresponding responses from the rotor servos. DASE is a command-augmentation system, so is designed to modify the control responses to match the pilot's demands. For example, if the pilot makes a sudden demand via the cyclic stick for a change in attitude, the DASE initially translates this into a crisp response whose magnitude is then reduced as the manoeuvre is carried out. To the pilot, the result is a responsive and natural-feeling movement of the aircraft.

At airspeeds of above 60kt, DASE also acts as a turn coordinator. 'Normally, in balanced flight a single-rotor helicopter with a tail rotor has about four degrees of right sideslip,' explains Steve Hanvey. 'Whatever sideslip angle you trim to, DASE will then maintain that. As you go into a turn if will sense the sideslip and maintain it fairly close.'

Electrical sensors continuously monitor the movements of the pilot's con-

Left: The pilot's cockpit has the clean and uncluttered appearance which results from modern avionics and display technology. The centrally mounted Video Display Unit can show either TADS or PNVS imagery as well as basic flight information.

Above: Apache gunners are classified as Co-Pilot/Gunners (CPGs). In addition to the TADS controls to the left of the displays, they also have basic flying instruments (right), and a centrally mounted control stick so they can fly the helicopter.

trols and the responses of the rotor servos, ensuring that all control demands are being acted on. If mechanical control is lost in any axis as a result of mechanical failure or combat damage, the lack of response is immediately noticed, and a fly-by-wire Back-Up Control System (BUCS) within the DASE takes over the affected control axis. The astute reader may wonder what happens in the event of the controls jamming, thus immobilising the control column and preventing the pilot sending manoeuvre demands

Below: This view of Apache shows the under-fuselage blade antennas for the VHF radio, UHF radio, and radar-warning receiver. The box-like objects on the tail boom are dispensers for chaff and IR flares; the ALQ-144 IR jammer is behind the rotor mast.

for the BUCS. No problem – a sharp movement will allow the pilot to break a shear pin at the base of the stick, disconnecting it from the jammed mechanical system.

When Apache entered production, testing of the back-up control system had not reached the point where the production configuration could be verified. At a time when the company and the Army were working to eliminate the inevitable problems experienced with any complex new aircraft, work on the fly-by-wire system was assigned a low priority.

An incident involving PV31 led to an Army review of the back-up control system. Although the equipment was installed on early production aircraft, it was mechanically and electrically disconnected. McDonnell Douglas studied the Army's recommendations, and by the early summer of 1986 was preparing a revised flight-test programme under which the system would be proven and available for Army use by the end of the year.

Primary navaid is the Singer Kearfott ASN-128 Lightweight Doppler Navigation System (LDNS). Developed as a replacement for the earlier ASN-64A, the LDNS is a completely self-contained navaid able to provide accurate aircraft velocity, present position and steering information from ground level to well above 10,000ft (3,000m). It consists of an RT-1193 receiver/transmitter/antenna and a CV-3338 signal data converter (which together form the Doppler Radar Velocity Sensor), plus a CP-1252 computer display unit and an optional steering hover indicator unit (SHIU) for the pilot's use. The complete installation weighs 28lb (12.7kg), consumes just under 90 watts of electrical power and has a MTBF (Mean Time Between Failure) of 2,121hr.

Operating frequency is in K-band (20-40GHz), with four FM/CW (Frequency-modulated/Continuous-Wave) beams being directed towards the terrain below the aircraft from a single antenna. The returned signal is received by the same antenna, then processed to extract velocity data. Given data on aircraft heading and altitude, the system can calculate aircraft position and velocity and provide the pilot with steering information.

More than 1,700 ASN-128 systems have been delivered for use on various US Army aircraft, and it is built under licence by Standard Elektrik Lorenz in West Germany for use in antitank and liaison/observation helicopters of the West German armed forces. Other Apache navaids include the Litton ASN-143 Heading and Attitude Reference Set and an Emerson Electric ARN-89B radio direction-finder set.

EW equipment

Like their fixed-wing counterparts, US attack helicopters now carry electronic warfare equipment intended to detect hostile radars and to confuse radar and infra-red guided weapons. In equipping Apache, the US Army turned to systems already developed for the earlier AH-1 Hueycobra.

Apache is fitted with the E-Systems APR-39 radar warning receiver. This lightweight unit weighs only 8lb (3.6kg) without its associated cables and mounting brackets. Used by all three US Services, it is installed on the UH-1 Iroquois, OH-6 Cayuse, OH-58 Kiowa, CH-46 Sea Knight, CH-47 Chinook, OV-1 Mohawk and OV-10 Bronco. In foreign service it equips types such as the BO105, Lynx and Gazelle, as well as the Hawker Hunter fighter.

An underfuselage blade antenna plus four spiral antennas monitor signals in E, F, G, H and I band (2-10GHz), plus parts of the 500MHz-2GHz C and D bands, and J-band frequencies above 10GHz. Warning of detected threats is given by alarm tones which sound in the aircrews, headsets. The ALR-39 displays the bearing of the emitter on a 3in (7.5cm) diameter CRT indicator, while an indication of the approximate pulse-repetition frequency is given by an audio tone.

An improved version of the APR-39 has now been developed. Known as the -39A, this can detect signals from C to M band (500MHz-100GHz) and locate the

bearing of signals in bands E to M (2-100Hz). Interfaces are provided for add-on sensors such as millimetre-wave and laser detectors, or missile-warning devices. A batch of 23 systems was assembled for the US Army and Marine Corps in 1984, but there are no reports of its being fitted to Apache.

Laser warning

Space has been provided within the aircraft for the future installation of a Perkin-Elmer AVR-2 laser-warning receiver. Designed for US Army and Marine Corps use as a complement to the APR-39, it was successfully flight tested at Fort Knox in the early 1980s, and is now in production. It is able to detect laser energy reaching the aircraft from anywhere within its 360° field of view, identify the type of threat, then use the audio and visual indicators of the APR-39 to warn the crew.

Another item of AH-1 hardware to have found its way into Apache is the ITT Avionics ALQ-136 radar jammer. This internally mounted system weighs 42lb (19kg) and consists of an LRU-1 transmitter/receiver/signal processor, a flush-mounted spiral LRU-2A receive antenna plus a similar LRU-2B transmit antenna, and a cockpit-mounted LRU-3 control unit. Operating frequency is in I/J band, and the equipment is understood to use angle and range deception techniques. It is probably intended to confuse the J-band Gun Dish radar carried by the ZSU-23-4 Shilka self-propelled anti-aircraft gun, and SA-9 Gaskin missiles during pop-up manoeuvres. When a hostile signal is detected, the unit analyses the received signals, then energises its transmitter to provide the appropriate jamming signal.

Infra-red threats such as heat-seeking air-to-air and surface-to-air missiles are the target of the Sanders Associates ALQ-144 infra-red jammer. Already in service on the UH-1 and AH-1, this uses a cylindrical ceramic block heated by electrical power as a source of infra-red energy. This is surrounded by a modulation system which causes the IR energy to vary in a pattern designed to confuse missile seekers by creating false error signals.

A chaff dispenser is also fitted. This is probably the Tracor M130, a US Army development of the ALE-40. The latter is in widespread service, and is carried by fixed-wing aircraft such as the F-5E/F, Hunter and Mirage. Primarily intended to launch M-1 chaff cartridges, it can also be used to dispense M-206 IR flares.

Three electronics units make up the Apache communications system. Voice links between the pilot and gunner are handled by the Telephonics C-10414 pilot communications system, with VHF and UHF radios being carried to maintain links with other aircraft, ground forces, and its own base.

Communication system

VHF communications are provided by the Collins ARC-186 tactical VHF radio. Currently the standard USAF communications equipment for airborne use at these frequencies, this set covers frequencies from 20 to 88MHz using AM (Amplitude Modulation), and from 108 to 152MHz in FM (Frequency Modulation) mode. (Readers who wonder why the 88-108MHz range is not covered will find the answer displayed on the dial of their own FM radio receivers – these frequencies are assigned to broadcast use). Up to 20 channels may be programmed for pre-selection during flight, while two dedicated selector switches cover the 40.5MHz and 121.5MHz 'guard' channels used for emergency communications. Maximum output power of the -186 transmitter is 10 watts. UHF coverage between 225 and 400MHz is provided by the Magnavox ARC-164. Like the VHF equipment, it offers 20 pre-selected channels and 10 watts of transmitter power.

One final element of communications-related equipment is a Bendix APX-100 transponder. When this receives an interrogation signal from ground or airborne IFF systems, it transmits a coded reply which will identify Apache as 'friendly' on NATO radar screens – a vital facility in combat.

The Apache avionics suite also includes Built-In Test Equipment (BITE). In Apache, this is known as the Fault Detection and Location System, or less formally as 'Fiddle'. The TADS, IHADSS, Doppler navigation and fault detection/location subsystems are all interconnected by a multiplexed MIL-STD-1553 digital databus, a single twisted pair running along either side of the helicopter replacing conventional wiring looms. Back in 1978 Army Project Manager Brig Gen E. M. Browne estimated that these two twisted pairs played the same role as 'more than 150ft (45m) of wire bundles each 3.5in (9cm) thick'.

Pilot's cockpit display panel

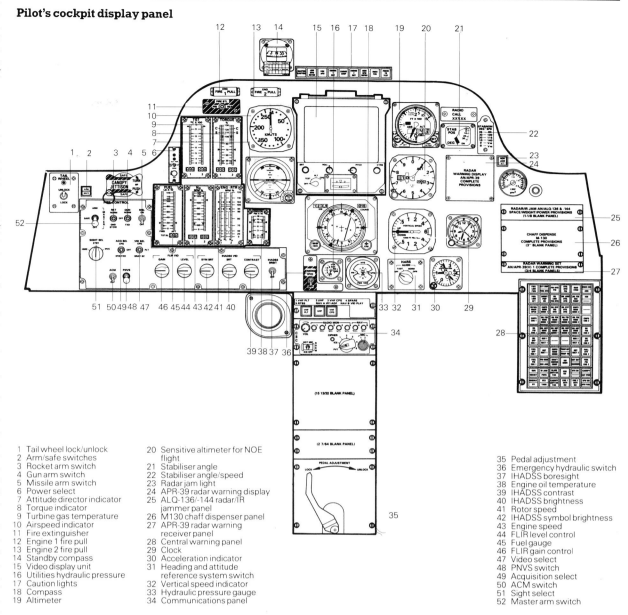

1 Tail wheel lock/unlock
2 Arm/safe switches
3 Rocket arm switch
4 Gun arm switch
5 Missile arm switch
6 Power select
7 Attitude director indicator
8 Torque indicator
9 Turbine gas temperature
10 Airspeed indicator
11 Fire extinguisher
12 Engine 1 fire pull
13 Engine 2 fire pull
14 Standby compass
15 Video display unit
16 Utilities hydraulic pressure
17 Caution lights
18 Compass
19 Altimeter
20 Sensitive altimeter for NOE flight
21 Stabiliser angle
22 Stabiliser angle/speed
23 Radar jam light
24 APR-39 radar warning display
25 ALQ-136/-144 radar/IR jammer panel
26 M130 chaff dispenser panel
27 APR-39 radar warning receiver panel
28 Central warning panel
29 Clock
30 Acceleration indicator
31 Heading and attitude reference system switch
32 Vertical speed indicator
33 Hydraulic pressure gauge
34 Communications panel
35 Pedal adjustment
36 Emergency hydraulic switch
37 IHADSS boresight
38 Engine oil temperature
39 IHADSS contrast
40 IHADSS brightness
41 Rotor speed
42 IHADSS symbol brightness
43 Engine speed
44 FLIR level control
45 Fuel gauge
46 FLIR gain control
47 Video select
48 PNVS switch
49 Acquisition select
50 ACM switch
51 Sight select
52 Master arm switch

Co-pilot/gunner's cockpit display panel

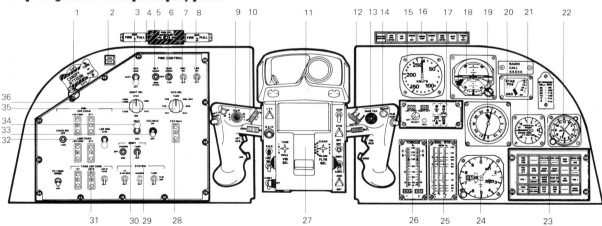

1 Canopy jettison
2 Jettison arming switch
3 Weapon arming switch
4 Fire switches
5 Rocket normal/off/ground stow
6 Gun normal/off/fixed
7 Missile on/off
8 Laser on/off
9 Weapon select
10 Sensor select
11 Sensor eyepieces
12 Laser tracker on/off/manual
13 Manual tracking control
14 Slave switch
15 Airspeed indicator
16 Caution lights
17 Engine and fuel status
18 Attitude director indicator
19 Compass
20 Stabiliser angle
21 Vertical speed indicator
22 Clock
23 Warning light panel
24 Altimeter
25 Engine and rotor speed
26 Torque indicator
27 Video display screen
28 Targeting/navigation select
29 Targeting system selectors
30 Boresight selectors
31 TADS laser code
32 Missile laser channel select, upper and lower channel code and quantity
33 Fire control computer multiplex
34 Multiplexer
35 Acquisition system selector
36 Sight selector

Armament

Anti-tank missiles have in the past been slow-flying weapons trailing the guidance wires along which steering command were sent. Operators of first-generation weapons such as SS.11 and AT-3 Sagger must often dream of how convenient a homing weapon would be. Most existing anti-tank helicopters are firmly tied to their missiles by guidance wires, and are thus limited in stand-off range. For Apache, Rockwell International created Hellfire – a homing missile which out-ranges wire-guided types – while new versions of the proven 2.75in unguided rocket were also developed, along with the novel Chain Gun 30mm cannon.

Apache has four weapon stations – two under each wing. Location of the wings is near the aircraft centre-of-gravity, so that the installation of ordnance results in no major shift of the c.g. location. The aircraft has sufficient performance to carry more than 5,000lb (2,270kg) of payload, but at present the Army has no weapons capable of fully utilising this, so that with all hardpoints fully loaded Apache still has performance to spare.

Primary air-to-ground weapon carried by Apache is the AGM-114A Hellfire anti-tank missile. Targets may also be engaged using 2.75in unguided rockets or the built-in 30mm cannon, but the aircraft's main operational role is to carry out Hellfire attacks against armoured formations.

Hellfire has a complex history, having evolved from the earlier ZAGM-64A Hornet programme. The story began long before the AAH programme was first conceived, and spans more than two decades from original concept to first deployment. Work on an advanced air-launched anti-tank missile first started in 1963, a time when the only helicopter-mounted anti-tank missiles in Western service were primitive French SS-10 wire-guided weapons. At that early date only five years had passed since the US Army had abandoned Dart, its first home-grown anti-tank missile, and two more years would pass before Hughes started work on TOW.

Work on weapons of this type was triggered off by a North American Rockwell private-venture proposal designated ATGAR (Anti-Tank Guided Aircraft Rocket) which attracted Service interest and gave rise to the Hornet programme, an ambitious project intended to create a smart anti-tank missile able to home in onto its target, eliminating the need for trailing guidance wires and post-launch steering by the missile operator. Rockwell set out to examine the feasibility of TV and electro-optical homing systems.

Hornet abandoned

Test firings started in 1964. Hornet was similar in external appearance to today's Hellfire, but was longer and lacked the latter's canard surfaces: 6ft 6in (1.98m) in length, and 7in (18cm) in diameter, it weighed 125lb (57kg) and was usually flown with a vidicon TV fire-and-forget guidance system. Attractive though the concept may have seemed on paper, the missile was abandoned as a weapon in March 1968. True fire-and-forget performance – the packaging of Maverick-style guidance into a missile only one fifth of the weight – was probably beyond the technology of the time.

With Hornet dead, the US Army looked at alternative weapons able to meet its requirement for an air-launched missile. All were derivatives of existing hardware, and while the obvious starting point was Hornet, there were other candidates. General Dynamics offered MRAM (Multi-mission Redeye Air-launched Missile), an adaptation of its man-portable light SAM. Philco Ford's MGM-51A Shillelagh gun-launched anti-tank missile was already in US Army service as armament of the M551 Sheridan light tank, and the company proposed a helicopter-launched adaptation aimed via a gyro-stabilised sight.

The basic concept of the Hellfire (HELicopter Launched FIRE-and-forget) missile was drawn up in 1970, and Hornet was reactivated to serve as a testbed for trials of seekers for future missiles. By this time the US Army was firmly wedded to the concept of anti-tank helicopters, and was working on an air-launched version of the TOW missile. First firing of TOW from a helicopter took place in July 1970 when a Lockheed Cheyenne fired the weapon from a three-round pack, scoring a hit on an M4 Sherman tank target, and in March of the following year Bell was given a $24 million contract to begin the task of adding TOW missiles to the AH-1 Huey-Cobra.

Development of Hellfire started in 1971, and test firings using Hornet began in that year. For these experiments Hornet was fitted with a more powerful rocket motor and carried various types of laser and electro-optical seeker. In 1971/72, the US Army's Missile Command carried out a very successful series of tests using Hornet airframes fitted with laser guidance. This seemed a promising approach, but the DoD was aware of many potential problems.

Laser drawbacks

For a start, laser seekers would not give true fire-and-forget capability. The target needed to be designated by a ground or airborne laser, perhaps not for the entire duration of missile flight, but certainly for the final portion. Poor weather could cut visibility to the point where range would be drastically reduced. The ability to fire weapons in rapid succession at multiple targets was also desirable: could a laser be switched rapidly enough from a newly destroyed target to a fresh one so that the next

Above: No public relations department can resist the temptation of taking this type of "what we can carry" photo, despite the fact that no aircraft could simultaneously carry all the stores arrayed before it.

Left: An AH-64 prototype poses with a dummy AGM-114A Hellfire missile installation. Note the movable trailing edge surface on the stub wing, a feature dropped from production aircraft.

Rockwell AGM-114A Hellfire

Hellfire is guided by a semi-active laser seeker in the extreme nose. When the round reaches its target, the shaped-charge warhead aft of the canard control fins is detonated. The resulting jet of hot gas is designed to penetrate tank armour, so slices effortlessly through the missile guidance section and seeker.

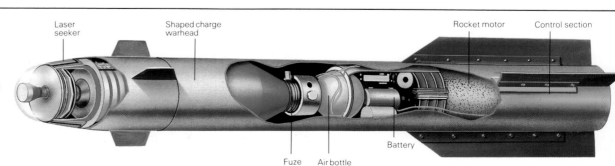

Hellfire laser seeker

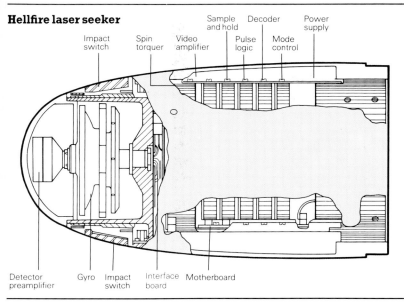

Left: Much research was carried out into alternative patterns of seeker for Hellfire. The most sophisticated designs such as imaging infra-red, advanced TV, and even dual-mode RF/IR homing heads would have given true fire-and-forget capability, but the US Army eventually opted for this Martin-Marietta semi-active laser seeker. As a result, targets must be designated with laser energy throughout the missile's flight.

Below: An Apache prototype fires a Hellfire missile. Operational testing of the weapon took place in 1980, and problems delayed the start of production until 1982. Trials have shown that the weapon can engage targets by day or night, scoring an impressive kill rate even in the presence of smoke and dust. Warsaw pact planners are doubtless wondering just how effective the weapon will be in a European winter.

Above: This trials film shows just how effective Hellfire is when faced with the armour of a medium battle tank. In this sequence, the laser-guided round can be seen striking an M48 Patton tank. The armour is successfully penetrated, causing the damaged vehicle to "brew up".

round in the salvo would receive guidance signals? Given these problems, the attractions of true fire-and-forget systems still beckoned.

When the US Army ordered prototypes of the YAH-63 and -64 in June 1973, the armament envisaged was still TOW, but advanced development of Hellfire was also under way, and selection of a definitive seeker was to drag on for some time. Laser Hornet had originally been fitted with a Rockwell-developed seeker based on the Sidewinder homing head. This was a logical approach to the creation of a low-cost seeker able to operate at wavelengths not too much shorter than those the unit had been originally intended to handle. Unfortunately, flight trials showed that this Sidewinder-derived unit had too narrow a field of view, so a new seeker with a wider angular coverage was developed and tested. This was seen as a possible tri-service design which in addition to being fitted to the Army's Hellfire could also be used on the USAF's Maverick and the US Navy's Bulldog, the last being a smart version of Bullpup which entered development in 1969 but was never adopted for service.

Hellfire contracts

In June 1974 Rockwell and Hughes were awarded competitive contracts for advanced development of what was termed the Hellfire Modular Missile System (HMMS), an airframe capable of carrying various types of seeker, and several patterns of advanced seeker were duly developed and tested. Advanced TV Seeker (ATVS) was a low-cost unit for daylight use only, and was developed in competitive forms by Hughes and Martin Orlando: one sixth the weight and half the diameter of the Maverick TV seeker, it could track smaller targets at longer ranges and in poorer image contrast conditions than similar units already in production. The IRIS (Infra-Red Imaging Seeker), on the other hand, was for day or night use: carry trials of rival Hughes and Texas Instruments designs started in 1973, and an improved version intended to have better performance in bad weather was subsequently developed by Hughes. Both were true fire-and-forget systems. Once the seeker had been locked on to its target by positioning a tracking gate around the target image on a cockpit-mounted CRT display, no post-launch action by the gunner was needed.

The breakthrough which allowed Hellfire to break free from this morass of competing seekers occurred in the same year, when US Army crews carried out a dozen firings of Hornet missiles equipped with a Rockwell-designed laser seeker. Launched at the Redstone Arsenal test range from AH-1G helicopters, these demonstrated the accuracy, range, and performance envelope likely to be achieved by Hellfire.

Trials carried out in late 1974 also tested an aspect of performance which would be crucial to the AAH, namely the ability of a single helicopter to ripple fire rounds at several targets. Two rounds

Above: First launch from Apache of a production-standard Hellfire missile was carried out at Yuma, Arizona. During the same series of tests, company and Army test pilots also fired missiles at night, and against targets located by the OH-58D AHIP scout helicopter. Any lingering doubts over Hellfire's performance under most weather conditions were dispelled by these tests, which cleared the way for aircraft and missile to enter US Army service. Just how well the weapon will cope with European winter conditions remains to be seen.

were fired eight seconds apart against two tank targets. A ground-based laser designator was used to illuminate the first tank, providing the energy onto which the first missile homed, scoring a direct hit. The laser designator was then quickly realigned with the second target, allowing the second missile to acquire its target.

Advanced trials

Two weeks later an even more difficult test was staged. An AH-1G carried out a pop-up manoeuvre from ground cover, hovered at a height of 80ft (24m), and fired two rounds in quick succession at two tank targets stationed around 65ft (20m) apart. The first was illuminated by laser energy from a ground-based laser designator, while the second was illuminated by a laser carried on a second AH-1G. The two missiles impacted almost simultaneously on their respective targets.

Other trials conducted in 1974 involved AH-1G helicopters modified to the Hellfire configuration and tested at the Hunter Liggett Military Reservation. Two were test-flown with Rockwell AAS-2 Airborne Laser Tracker (ALT) systems designed to detect reflected laser energy within 60° either side of the aircraft nose, displaying target position on a cockpit-mounted HUD. Two more aircraft carried Ford Aeronutronic Airborne Laser Locator/Designator (ALLD) pods on the starboard stub wing, allowing the aircraft to designate its own targets for three Hellfires mounted under the port wing. The remainder carried six Hellfires, three on each side.

No launch was carried out at Hunter Liggett, but the carry trials were more successful than had been expected. Helicopters were able to detect up to five times the number of targets that had been anticipated, and simulated launch operations showed that total exposure times from target acquisition to missile launch were often less than 10 seconds. At last, solid evidence of the tactical usefulness of laser seekers existed, guaranteeing a useful weapon even if the more complex designs bogged down in technical problems or turned out to be too expensive.

Dual-mode seekers

One concept which fascinated the US Army was that of creating a dual-mode seeker able to home on either radar or infra-red emissions. Installation of such a unit on Hellfire's modular airframe would create a missile able to counter two Soviet air-defence systems – the SA-6 Gainful surface-to-air missile and the ZSU-23-4 Shilka self-propelled anti-aircraft gun. In the hands of the Egyptian Army, these had been credited with inflicting much of the mauling received by the Israeli Air Force in the opening stages of the October 1973 Middle East War.

Advanced development of a modular dual-mode Air Defense Suppression Missile (ADSM) was begun in November 1974 with Martin Orlando and Rockwell building Hornet test rounds fitted with competing designs of seeker. Operating in RF mode, the ADSM seeker was intended to home in onto the signals from the associated Straight Flush and Gun Dish radar systems which tracked targets for the SA-6 and ZSU-23 respectively. If these were shut down, the seeker was to switch to IR mode, homing onto the heat created by Straight Flush's electronics or the hot gun barrels of Shilka's quad 23mm cannon.

By early 1976 YAH-63 and -64 prototypes had been flying for 10 months, and Hellfire development had been going smoothly enough to allow the decision to be taken on February 26 to substitute Hellfire for TOW. Rockwell was selected to develop the definitive missile, receiving an engineering development contract in October 1976, and two months later Hughes was named winner of the AAH competition.

The eventual choice for initial rounds was semi-active laser guidance, and the Army intended to fit the missile with the Rockwell Tri-Service laser seeker, a homing head also planned for use on the AGM-65C version of Maverick and the GBU-15 glide bomb. Work on an imaging infra-red seeker continued for some time, but was eventually cancelled in February 1981.

Despite the obvious advantages of employing a seeker of standardised and common pattern, studies suggested that a simpler and cheaper design of homing head would be better for most Hellfire shots, which would involve short-range engagements of multiple targets. Martin Marietta is reported to have offered the Army an alternative pattern of low-cost seeker which it had developed as a private venture. Following a design competition against Rockwell, the Army selected Martin Marietta to develop the definitive Hellfire seeker.

Hellfire is smaller and lighter than Hornet. The missile is 5ft 4in (1.62m) in

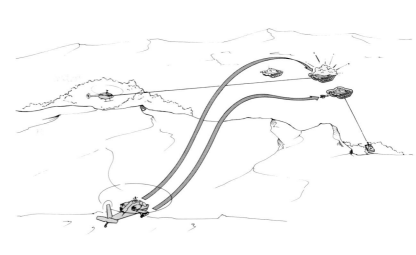

Hellfire indirect launch

Above: US Army crewmen load Hellfire missiles onto Apache's launch rails. Each launcher is designed to carry a total of four rounds. Given its full load of four launchers, each Apache could in theory dispose of up to 16 enemy armoured vehicles.

length, 7in (17.8cm) in diameter, and has a tail wing span of 12.8in (32.6cm); in its laser-homing form it weighs 95lb (43kg). The Cassegrain optical system of the seeker is mounted immediately behind a window at the front of the nose section, a cylindrical assembly which also contains the seeker electronics, and has four fixed canard surfaces at its aft end. Next in line is the warhead section, which contains the 20lb (9kg) shaped-charge warhead and the fuze. Then comes the guidance section containing a pneumatic accumulator, pitch and yaw/roll gyros, autopilot electronics and the battery. The final section contains the solid-propellant rocket motor and four long-chord wings which terminate in control surfaces.

Firings from a JAH-1G Cobra trials aircraft started on October 21, 1978, and the first airborne guided shot in the Hellfire engineering development phase took place on December 17. Trials using the YAH-64 began with a series of unguided test shots, which led to the first guided Hellfire launch from this aircraft on September 18, 1979. Operational Testing (OT) II trials were carried out between April and July of 1980, and development testing ended in December of the same year.

Hellfire launchers

Apache carries Hellfire on four wing-mounted modular launchers located inboard and outboard under the aircraft's stub wings. Each launcher has four rails, but the lower pair may be detached if necessary.

Like many US weapon systems, Hellfire had been developed using test ranges whose weather conditions were much better than those which the missile would face when deployed in Western Europe. Congress had expressed concern on this account, so in 1981 the Army fired eight missiles in conditions of artificially created fog, smoke, dust clouds and rain. Two shots failed as a result of the obscuring clouds reflecting the laser energy being used to illuminate the target, but the remaining four were successful. As a result of this experience, the Army admitted that the weapon would not be able to cope with some weather conditions, but stated that alternative tactics such as the use of ground-based laser designators would allow Hellfire to cope with most weather.

Ambush positions

Above: An aircraft as expensive as Apache will try to stay well back from hostile defences. Whenever possible it will fire against targets already located and designated by ground or airborne laser illuminators rather than go looking for them.

Left: Hellfire is light enough to be manhandled into position – no specialised loading aids are needed. This simplicity will pay dividends when operating from frontline locations: as long as supplies of fuel and ammunition are available, any open space can act as an airstrip.

Below: In combat, Apaches will try to move forward to secondary hiding places; once targets are within range the aircraft will take up their final ambush positions, then launch their missiles. Like a tank, Apache will fire a few rounds then withdraw to a new ambush position.

Hellfire was not completely free from teething troubles. In May 1981 several rounds malfunctioned during a series of 20 system-qualification test firings from a tower-mounted launcher. This was found to be the result of a gyro fault. Final pre-production testing took place in the summer of 1981 at Hunter Liggett Army Reservation, California.

Production go-ahead for Apache and Hellfire was originally expected to be linked, but in 1982 the DoD decided that the two programmes should be treated separately. On March 31 of that year, Hellfire was approved for production, which started with an 884-round buy in FY82. Missiles and launchers were rolling off the production line by the summer of 1984, as the Army was conducting final tests.

Multiple target designation

Hellfire has been tested successfully with rapid and ripple firings using two or more targets and designators. During a trial carried out at the Yuma Proving Ground in Arizona using production missiles, two rounds were launched at night on November 17, 1984, from a low-flying Apache. 'Both missiles hit and destroyed targets at ranges well beyond the specified maximum night engagement ranges,' the Pentagon later stated. Firings in which the target was initially designated by one designator and subsequently handed off to a second designator have also been successful.

Hellfire entered service on the AH-64 in January 1985 following delivery of the first 18 helicopters. (The Army inventory requirement for the missile is estimated at 47,100 missiles; the Navy also has indicated an inventory requirement estimated at about 11,000.) By that time, the weapon already incorporated some improvements. A low-smoke motor had been successfully developed and tested, along with an anti-icing shielded dome for the laser seeker and a sealable container for the complete missile.

In keeping with the present practice of buying missiles from two rival assembly lines, Hellfire has now been second sourced. During FY84 Rockwell passed the necessary data to Martin Marietta, and both companies now manufacture the weapon. By the end of 1984 Martin Marietta had produced its first Hellfire missiles, the initial shipment of what was to be a batch of 3,000 over the next three years.

Hellfire will also arm other types of helicopter, and has even been proposed as a weapon for fixed-wing and ground-launched applications: in 1984 Rockwell was asked by the US Navy and Marine Corps to carry out a six-month study of possible Hellfire armament for the OV-10 Bronco, AV-8B Harrier II, and F/A-18 Hornet. In November of the same year, the US Army's Aviation Systems Command gave Sikorsky a $7.9 million contract to qualify the Hellfire missile for use on the UH-60 Black Hawk. The Navy began modifying the Hellfire rocket igniter to make it compatible with shipboard use.

Hellfire applications

These were not the first studies of alternative Hellfire platforms. The missile had already carried out captive flight trials on the OV-10, and three rounds had been test-fired from a UH-60 in 1982 with no adverse effects on the aircraft or missile. When fitted with an External Stores Support System (a kit of removable pylons carried on the aircraft's sides), the UH-60 can carry Hellfires, mine dispensers, external tanks or other equipment. One obvious application was the AH-1, and in FY86 the US Marine Corps took delivery of 44 AH-1T Super Cobra helicopters. These are able to carry several types of missile, including TOW, Hellfire, and even Sidewinder.

Hellfire has also been suggested as armament for the Agusta A.129 Mangusta and the Westland Lynx 3; the latter can carry up to eight Hellfires. Firing trials were carried out at a Norwegian range in 1983 using a Ferranti 306 target designator, but no user of the aircraft has so far opted to procure the system.

Rockwell has also offered Hellfire as a vehicle-mounted ground weapon, and has carried out firings from a truck. The only customer to have adopted the weapon in this role by mid-1986 was Sweden, which intends to deploy Hellfire as a coast-defence missile. Late in 1984 Rockwell was given a $7.7 million two-year development contract for the new variant, which will be fired from a Swedish-designed single-rail launcher and storage container. Rockwell will act as systems integrator and will supply the missiles, but these will be fitted with a Swedish-designed warhead optimised for use against naval targets.

Hellfire is intended primarily for use against tanks and other fighting vehicles. For attacks on soft targets Apache crew can use unguided 2.75in folding-fin rockets. Unguided rockets have been fired from US aircraft since World War II, and development of the first 2.75in rockets took place in the immediate postwar years. Weapons of this calibre were a standard form of fighter armament in the pre-missile era, arming veteran types such as the F-86D and -86L all-weather versions of the Sabre, and subsonic interceptors such as the F-89 Scorpion and F-94 Starfire. They remained a feature of the armament of the F-102 Delta Dagger and F-106 Delta Dart, but quickly gave way to radar-guided or IR-homing missiles.

Improved rockets

Despite its basic simplicity, the 2.75in rocket has been steadily improved. The US Navy developed a higher-thrust Mk 66 motor which has since been adopted for tri-service use. Army procurement is now concentrated on the new Hydra 70 weapon system, which combines the Mk 66 motor with new launchers and warheads. These weapons will arm the AH-64, as well as the AH-1J and OV-10 Bronco.

Left: Not all targets will be worth an expensive Hellfire round, so Apache can also carry lightweight M260 and M261 launchers for 2.75in unguided rockets. This aircraft is carrying two 19-round M261 launchers.

BEI Hydra 70 unguided rocket warheads

M261 multi-purpose submunition

Right: The M261 multi-purpose submunition warhead has an M439C fuze and carries high-explosive shaped-charge munitions for anti-armour, anti-materiel and anti-personnel use.

M247 shaped charge

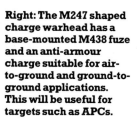

Right: The M247 shaped charge warhead has a base-mounted M438 fuze and an anti-armour charge suitable for air-to-ground and ground-to-ground applications. This will be useful for targets such as APCs.

M255 flechette

Right: Payload of the M255 warhead consists of around 2,500 flechettes, each weighing 28 grains (430 grams). Released by the operation of an M439 fuze, these are effective against personnel and materiel.

M264 smokescreen

Right: Obscuration and target marking are the roles of the M264 smoke screen warhead. Triggered by a base-mounted M439 fuze, this releases a cloud of smoke which persists for up to five minutes.

Armament

Above: Apache prototype AV04 ripple-fires a series of 2.75in rockets. The older Mk 40 rocket is being replaced by the improved Hydra 70 series, which incorporate the new Mk 66 pattern of rocket motor.

Right: Rocket-firing tests at the Yuma Proving Ground. Hydra 70 rounds are designed to spin up in the launcher in order to obtain maximum accuracy. Range of these new rockets is up to 9,600 yards (8,800m).

For the Apache and AH-1 Cobra, Hughes Aircraft manufactures light-weight aluminium launchers of seven-round and 19-round capacity. Developed under a contract from the US Army Missile Command's Redstone Arsenal, the units are assembled by pressing rather than by welding or riveting, a technique which minimises weight and reduces labour costs. Although inexpensive enough to be disposable, they are strong enough to withstand extended use, and during trials two seven-round launchers were still fully functional after 65 firings.

Both patterns of launcher have proved reliable. Five experimental models of both sizes were used to discharge a total of 4,500 rockets during tests at the Army's Yuma Proving Grounds, Arizona. The trials combined shots from ground test mountings and for aircraft, and proved the concept and detailed design of the launchers.

A $4.3 million production contract for 1,185 M260 (seven-round) and 785 M261 (19 round) launchers was awarded in November 1979. First deliveries took place in the following summer, and at least one follow-on order has since been awarded, while Harbard Interiors has been set up as a second manufacturing source: in February 1986 the company was given a $3.5 million contract to build 10 prototype and 40 production M260 launchers, plus 10 prototypes and 1,860 production versions of the M261.

Length of the M260 and M261 is 5ft 5in (1.65m). The seven-round M260 is 10in (25.4cm) in diameter and weighs 34lb (15.5kg) when empty, while the 19-round M261 is 16in (40.6cm) in diameter and weighs 79lb (35.9kg). Both launchers can accept rockets fitted with Mk 40 or Mk 66 motors, and once rounds have been loaded into their individual tubes, a locking device holds them safely in place.

With an M261 launcher on each underwing station Apache can carry a maximum payload of 76 rounds. Rockets can be fired singly, in pairs or in four-round salvos, the interval between individual firings being 0.06 sec, and fuze timing can be remotely set from the aircraft cockpit. When the rocket is fired an electrical impulse from the firing switch reaches the rocket motor via a pivoting contact arm mounted at the after end of the launch tube; as the motor ignites the hot gases push the contact arm rear-

Below: Being relatively simple, unguided rockets are easy to store and handle; and being relatively cheap, they can be fired in large numbers. Annual orders placed by the US Department of Defense can vary from tens of thousands to more than 100,000. Simplest model in the Hydra 70 range is probably the version fitted with the M151 high-explosive warhead. The more complex versions are described opposite.

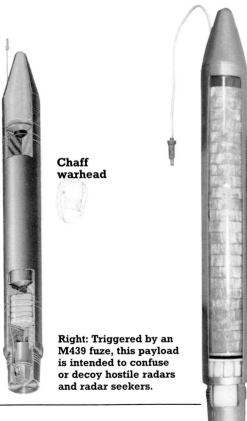

M262 illumination

Chaff warhead

Right: Given the capabilities of its night-vision systems, Apache should rarely need to fire the M262 illumination warhead. This is designed to provide one million candle power for approximately two minutes.

Right: Triggered by an M439 fuze, this payload is intended to confuse or decoy hostile radars and radar seekers.

ward, causing a mechanical connection to unlock the round and allowing it to leave the tube.

The 2.75in rocket is essentially an area-saturation weapon. It has a maximum useful range of 5.5km and can carry high-explosive or white phosphorus warheads fitted with a variety of fuze systems. Warheads are also the subject of regular upgrades and improvements, and a wide range available for use in the Hydra system.

The M259 smoke warhead was based on white phosphorus, but the newer M259E1 uses red phosphorus and an electronic fuze. Development of the M261 multipurpose submunition warhead with a remotely set fuze was completed in 1981, and there is a training version of this payload, the M267 smoke/flash warhead. The M262 illuminating round is similar to the earlier M257, but has an electronic fuze. Design of the M257 warhead has not yet been finalised, but the payload will consist of 18 flechettes.

Being relatively inexpensive, unguided rockets are purchased and used in large numbers. For example, the FY86 the US Army bought 23,000 M151, M423 and Mk 66 rockets at a cost of $7.0 million, and planned to spend $44.8 million the following year to buy another 185,000. Typical USMC annual spending on rockets in the mid-1980s was around $60 million.

Under the belly of the Apache is the M230 30mm Chain Gun. This single-barrel cannon introduces new levels of safety and reliability, yet the secret of its success does not involve exotic high technology – just a length of industrial chain as used in thousands of drive belts, plus the ingenuity of the weapon's designers.

Chain Gun principle

The basic principle of single-barrel machine guns and cannon is little

Above: During Operation Desert Storm, the Chain Gun showed its worth in combat conditions, proving both reliable and deadly against 'soft targets'.

changed since US inventor Hiram Maxim invented the first practical machine gun in 1884. Part of the energy released by the exploding cartridge, obtained either by making use of the otherwise-wasted recoil energy (the technique used by Maxim) or by extracting a small amount of the hot gases from the barrel, is used to unlock and open the breech, extract the spent case, load a fresh round, close and lock the breech, then release the firing pin. When the pin strikes the primer at the base of the cartridge case, the cartridge explodes, and the entire cycle is repeated.

Gas and recoil systems share a common weakness: if a round fails to fire when initiated, there is no energy available to remove it from the breech, substitute a fresh round, close the breech and fire the weapon, and given the universal validity of what engineers term Murphy's Law – 'If it can go wrong, it will go wrong' – stoppages in combat are inevitable.

On an infantry-crewed ground weapon, the operator simply operates the cocking handle, using muscle power to drive the breech mechanism through a single unloading/reloading cycle. World War I pilots used the same routine to cure stoppages in the armament of their primitive fighters, but with the arrival of monoplanes and remotely-located guns, the problem became more serious, requiring the installation of powered recocking devices. On the French 30mm DEFA aircraft cannon, for example, this takes the form of a single pyrotechnic cartridge which may be triggered in order to generate an impulse able to recock the gun. This is a single-shot system; should the attempt fail, or the gun jam a second time, it will remain jammed for the rest of the sortie.

The only effective way to create a reliable single-barrel weapon would be to use some form of power-driven breech, relying on a external source of power for loading and unloading. Probably the best modern example of this approach is the General Electric M61A1 cannon carried by most US fighters, each of whose rotating series of barrels is at a different part of the loading/unloading cycle, and which obtains the energy needed to turn the barrels and operate the individual breech mechanisms from an electric or hydraulic motor.

The basic principle of the Vulcan was first used in the Gatling Gun, devised in 1862 by Dr Richard Gatling, which used the energy from a hand crank to rotate its barrel assembly. During early development work on the Vulcan, engineers even took a Gatling gun, renovated it, then coupled it to an electric motor and test-fired the result in order to obtain design data for the new weapon.

Hughes research

In 1972 Hughes Helicopters engineers began an independent research and development programme with the aim of developing an externally powered single-barrel weapon able to fire the US Army's existing M50 20mm round. The company had already developed such a weapon in 7.62mm calibre known as EPAM (Externally-Powered Armor Machine gun), a dual-feed weapon which showed high reliability compared with conventional machine guns and encouraged the company to apply the same concept to a cannon.

Work on a 20mm weapon started, and a mock-up of a dual-feed 20mm weapon was created. Studies soon showed that a Gatling-type rotating breech using a drum/cam follower action would be too complex and heavy for use in a single-barrel gun. What was wanted was some way of converting the rotary power from an electric motor to the reciprocating motion needed for a conventional gun breech.

Most cannon use a bolt mechanism to seal the breech once a round has been chambered in the form of a cylindrical piece of steel whose tip enters the mouth of the breech immediately behind the rear face of the cartridge. One or more lugs positioned around the circumference of the bolt tip lock into slots cut in the breech when the bolt is turned through part of a revolution.

What the Hughes engineers wanted was a method of moving the bolt into the breech, holding it there while the round was fired, moving it smoothly backward again to extract the spent case, then holding it in the fully-withdrawn position for a few moments before beginning the cycle anew. The solution turned out to be a length of industrial chain of the type used to create drive belts for machinery.

Below and behind the breech of the Hughes gun, a continuous belt of chain is positioned in a horizontal rectangular track. Three large-diameter unpowered idler sprockets plus a fourth powered sprocket hold the chain in the desired rectangular pattern, with the long axis

M230 Chain Gun

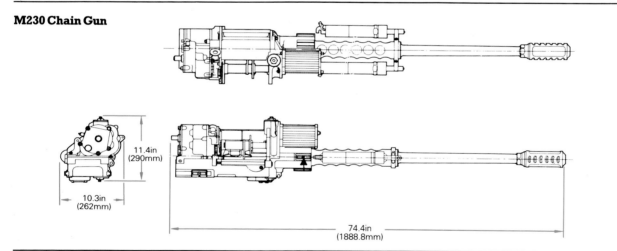

Above: The M230 Chain Gun is a compact single-barrel 30mm weapon devised by Hughes specifically for the AAH programme. Having defeated the rival General Electric XM188 gun it was adopted for service on Apache. Originally designed to fire a new pattern of US-designed ammunition, it was rechambered in 1976 to accept the ADEN/DEFA round already used by NATO.

Left: The simple lines of the Chain gun conceal a simple but novel mechanism. The breech of the weapon is driven by a 6.5hp motor via an endless belt of industrial chain which gives the weapon its name, plus its 15,000 mean rounds between failure level of reliability. Versions in 7.62mm calibre have been developed for other applications.

running parallel to the barrel. They also ensure that the corners of the rectangular pattern are well-rounded.

Bolt action

The bolt is in a T-shaped bolt carrier mounted in a horizontal position behind the breech and free to move forward and backward: the stem of the T houses the cylindrical bolt, while the cross-piece extends outward on either side for the full width of the rectangular loop of chain, and a slot cut in the lower surface of the cross-piece acts as a track for a drive shoe. The cross-piece, track and shoe are rather like a miniature version of our imaginary overhead crane.

The drive shoe is attached to a master link on the chain. As the chain rotates, the bolt carrier and bolt are dragged forward, toward the breech, pushing a round into the chamber. Continued movement forces the bolt face into the breech and provides the power needed to turn and lock it. As the master link travels along the short axis of its route, and the shoe slides along the length of the bolt carrier cross-piece, the bolt carrier and bolt remain stationary as the round is fired. The master link and drive shoe then move backward along the race track shaped route of the chain, pulling the bolt carrier behind them. The bolt unlocks and moves to the rear, extracting the spent case. Bolt and bolt carrier then pause at the opposite end of their travel, allowing time for the spent case to be removed and a new round to be moved into place between the bolt and the open breech.

Throughout the entire cycle the bolt is under positive control and subjected to

Right: Technicians learn how to maintain the Chain Gun steerable mounting. Reloading of the weapon is carried out from this location, with the endless carrier feed being used to deliver rounds to the magazine.

Chain Gun installation

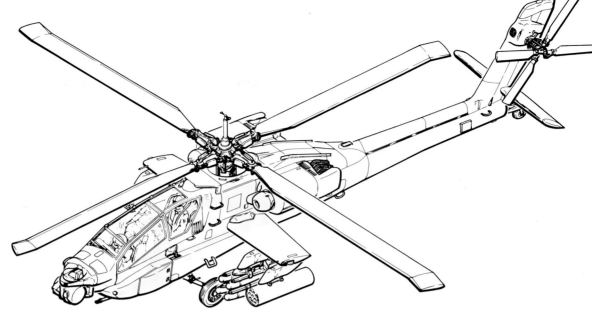

Above: Ammunition for the Chain Gun is stored in a 1,100 round "Flat Pack" magazine located close to the aircraft's centre of gravity. Rounds are not linked, but travel to the gun via an endless-belt carrier feed which runs down the starboard side of the fuselage and down to the gun. Spent cases are ejected overboard, and the now-empty feed returns to the magazine to be reloaded.

Below: Chain Gun is fired during a training sortie. The ADEN/DEFA round is old, but has sufficient destructive power for air-to-ground use. Its prime mission will be to deter enemy anti-aircraft systems.

low acceleration forces. The timing of the operating cycle is not dependent on variables such as weapon cleanliness, spring tension or gas pressure, and the breech is held firmly shut until gas pressure within the barrel has died away to ambient. The Chain Gun is simple, safe, resistant to the effects of dirt and wear, and very reliable.

Bigger calibre

By this time the US Army was looking for a 30mm weapon to arm the AAH. The AH-1 HueyCobra was armed with a 20mm weapon, and by switching to 30mm for Apache, the Army would gain extra range and hitting power. Hughes engineers noted that the proposed XM552/639 round which the new gun was to fire was similar in size and impulse to the 20mm M50, and realised that the design concept which they had devised could be applied to the Army requirement.

Development of what was then the XM230 started in December 1972. Work progressed rapidly, and the A Model prototype gun was fired for the first time in April 1973. Burst firing began in the following month, and in July Army Weapons Command sponsored a 2,500-round feasibility trial of the weapon. Minor problems were experienced, but by the time the test was completed on September 13, the A Model had fired the final 1,000 rounds without a stoppage or malfunction.

In January 1974 the improved B Model started a series of firing trials which would involve more than 50,000 rounds. This improved weapon had a simplified gearing arrangement, and a more precise indexing mechanism. Bursts of up to 300 rounds were fired at rates of up to 11 shots per second.

So far, all testing had been done with linked ammunition. This was convenient for trials purposes, but the US Army wanted a linkless feed. Having established confidence in the basic design, Hughes engineers were able to tackle the task of creating a linkless weapon, and the resulting C Model started firing trials in December 1974. Mounted in a prototype turret, it met the specified firing rate of between eight and 12 shots per second, and by the time it was fitted into a YAH-64 prototype, it had fired more than 100,000 rounds during tests at Culver City and at the Marine Corps base at Camp Pendleton, California. The aircraft, complete with gun and a 90-round linkless conveyor system, went through firing trials at Camp Pendleton and Edwards Air Force Base, discharging more than 2,000 rounds. Further ground testing at Edwards confirmed reliable operation in the presence of dust and sand.

The basic design was further refined to create the D Model, two examples of which were delivered to the US Army to begin a three-month competitive shoot-off against the General Electric XM188 at the Rock Island Arsenal. Both patterns of weapon were fitted with test instrumentation, then test-fired in rigid mountings and on test rigs intended to simulate aircraft. Firing began in March 1976 and lasted for three months. The rival guns were exposed to environmental stresses such as heat, cold and ice, and fired a total of more than 100,000 rounds. The US Army report which emerged quoted a mean time between failure (MTBF) of more than 12,000 rounds, and suggested that a figure of 50,000 hours would probably be possible after minor development testing.

On December 10, 1976, Hughes Helicopters was given the contract to supply both the AH-64A helicopter and its gun. By this time a quarter of a million rounds had been fired through the A, B, C, and D Models, but the development task was not yet over. Early in 1976 the DoD decided that the cannon of the AAH should fire a different round. The calibre remained the same at 30mm, but the ADEN/DEFA round widely used by the European NATO members was substituted for the US-developed ammunition. The European round was of relatively elderly design, but its performance was adequate for the task. More importantly, it was widely available from several production lines in Europe.

In March 1978 the rechambered gun, now known as the XM230E1, was being test-fired, and Hughes was developing a family of US rounds. Combat loads are the M789 high explosive dual-purpose and M799 high explosive, while the M788 is used for training.

The production M230 gun is 64.5in (163.8cm) long, 10in (25cm) wide and 11.5in (29cm) in height; total weight is 123lb (56kg). Maximum rate of fire is between 600 and 650 rounds per minute (around 10 rounds per second), and the barrel has a life of up to 10,000 rounds. On entering service, the gun had an MTBF of 15,000 rounds.

When the M230 is fired, a 6.5hp electric motor is energised, causing the endless chain loop to revolve in its guide. The gun takes 0.2 seconds to reach full rate of fire, and will cease firing within 0.1 seconds of the stop command.

Gun location

On the HueyCobra, the 20mm gun and its ammunition are located in the nose, underneath the gunner's station. As fuel is burned during the sortie and ammunition is expended, the aircraft centre of gravity is constantly changing. This affects the manoeuvrability of the aircraft. On Apache, the ammunition is located in a magazine positioned directly under the rotor mast near the aircraft centre of gravity. Referred to as a 'double flat pack', this two-story device carries 1,100 rounds of ammunition, well in excess of the 320 demanded by the US Army.

Rounds are not linked. A motor drive and transfer unit take the individual rounds out of the box and place them into a carrier drive which runs through a chute extending along both sides of the fuselage, then turn inward to meet above the gun. The rounds pass along the starboard side of the aircraft and down a flexible chute into the gun, where they are fired. The empty cases are ejected overboard, while the empty carrier drive passes up a second flexible chute before being routed back down the port side of the aircraft and returning to the magazine to collect fresh rounds. Capacity of the chute running from the magazine to the gun is around 90 rounds, so that the maximum load which Apache can carry is 1,200 rounds.

Ammunition is taken to the Apache in boxes containing linked rounds, and loading is carried out through the belly of the aircraft. One of the flexible chutes is disconnected from the gun, and connected to a ground support unit which takes the linked rounds from the storage box, strips off the links and loads the rounds into the chute, where they back-track to the magazine. When the bottom half of the magazine has been filled, the transfer unit then loads the upper section.

Contrary to common belief, Apache crews are unlikely to use the Chain Gun as a method of attacking soft-skinned targets. Early suggestions that the weapon might be used against Warsaw Pact mobile anti-aircraft systems such as the ZSU-23-4 Shilka self-propelled anti-aircraft gun are incorrect – Apache is too vital to the anti-armour mission to be spared for other uses. The most likely role of the M230 is that of defence suppression, keeping enemy heads down as the aircraft carries out its primary mission of killing tanks.

Future weapons

Although Apache already possesses formidable firepower, both the US Army and the manufacturer are anxious to enhance its combat capability, especially in the air-to-air arena, for it is now acknowledged that other helicopters – most notably the Mi-24 Hind, Mi-28 Havoc and Ka-41 Hokum – could well present a serious threat. Various trials have already been conducted and more are due to take place so as to provide the AH-64A with the potential to defend itself.

Some effort is being directed towards improving the M230 Chain Gun's potential in the air-to-air role, but the majority of these trials centre around missile armament. One of the first projects involved the well-proven AIM-9 Sidewinder, two missiles of this type being successfully fired from an AH-64A over the Army's White Sands test range in November 1987. Further Sidewinder launches have taken place since then but it appears that the most favoured AAM is a suitably modified version of the Army's man-portable, shoulder-launched General dynamics FIM-92A Stinger.

Work on integrating this weapon began in October 1987 and it too has been successfully launched from Apache, which completed a series of five test firings at Yuma in early 1989. Like Sidewinder, it is an IR-homer, but its greater degree of compactness means that Apache is able to carry two vertically stacked launch tubes on each wing-tip. More tests were scheduled for early 1990 and it now appears that Stinger will be added to the Apache's arsenal once the upgraded AH-64B and AH-64C models enter service.

Other types of missile have also been fired from the AH-64A although there is no guarantee that any of these will become standard armament in years to come. The Sidearm anti-radiation variation on the Sidewinder theme is one such weapon and this was successfully tested at the China Lake Naval Weapons Center on April 25, 1988 when it scored a hit on an armoured vehicle that contained an RF emitter.

The Shorts Helstreak helicopter-launched version of the Starstreak close-combat, high-velocity, ground-launched missile is another contender. Captive-carry trials were due to take place in 1990, with live firings to follow in 1991, and this laser-guided weapon may well figure in the armament of the Apache if it is eventually purchased for service with Great Britain's Army Air Corps. As with Stinger, Helstreak would be carried in two vertically stacked tube launchers mounted on the wing-tips.

Finally, there has been a Hughes-originated proposal to consider the AGM-65 Maverick ASM as an alternative to Hellfire in particularly high-threat areas where Maverick's stand-off capability could well come in useful. A maximum of four missiles could be carried, these using individual infra-red seeker heads slaved to the Apache's TADS, although the Hughes Missile Systems group is also developing an autonomous millimetre wave seeker that would be compatible with Longbow Apache. However, cost considerations and the lack of a definite US Army requirement seem likely to conspire against this proposal.

Apache stores options

Below: For the moment, Apache's only external stores are rocket launchers, missiles and fuel tanks. As service experience grows, other weapons could find their way on to the stub wings.

Right: Hellfire is expensive, so drill rounds are often carried during training missions. This Apache has live rounds on the upper launch rails and red-marked mock-ups on the lower positions.

1. 230US gal (871lit) long-range tank
2. Seven-tube rocket launcher
3. 19-tube rocket launcher
4. BEI 2.75in Hydra 70 rockets
5. 5in Zuni rocket
6. McDonnell Douglas AGM-84 Harpoon anti-ship missile
7. Rockwell AGM-114A Hellfire anti-tank missiles on four-rail modular launcher
8. McDonnell Douglas M230 30mm Chain Gun
9. Hughes BGM-71 TOW anti-tank missile and quad launcher
10. Kongsberg Penguin anti-ship missile
11. General Dynamics Stinger air-to-air missile and twin box launcher
12. AGM-122A Sidearm anti-radar missile
13. AIM-9L Sidewinder air-to-air missile
14. M130 chaff/flare dispenser
15. Sanders ALQ-144 infra-red jammer

Armament

Deployment

In Killean, Texas, slick-talking auto salesmen will be only too glad to help you trade in your old wheels for the latest Ford, Chrysler or Toyota, but at Fort Hood, just outside of town, a more significant trade-in is under way as experienced AH-1 HueyCobra crews swap their veteran AH-1s for the Apache. All planned Apache units will train there before returning to their home base, or heading across the Atlantic to West Germany and front-line deployments. Even the National Guard is getting in on the act as weekend warriors hand in their Cobras and move up to America's newest and deadliest attack helicopter.

The initial task of training Army personnel to operate the Apache was assigned to three bases of the Service's Training and Doctrine Command (TRADOC) – Fort Eustis in Virginia, Fort Rucker in Alabama and Fort Gordon in Georgia. By the end of January 1985 the US Army had taken delivery of its first three production Apaches, two having been delivered to Fort Eustis and the third to Fort Rucker. By the end of the year Fort Rucker was due to take delivery of 33 aircraft, with Fort Eustis receiving a total of 14.

By the time these first deliveries were made, production aircraft had flown around 10,000 hours. This figure rapidly increased as battalions formed at Fort Hood, and by the end of May 1986 production aircraft had clocked up 15,000 hours of flight time. By the year end, the company expected this figure to pass the 50,000 hour mark.

The prototype fleet eventually flew almost 5,000 hours in total. Although PV02 and 03 had gone to museums at Fort Eustis and Fort Rucker, life was not yet over for the remaining prototypes, and PV05 and 06 were still at Mesa. PV06 last flew in autumn of 1984, but remained at the McDonnell Douglas Helicopter plant. No decision on its future had been taken by the time the author visited Mesa in the summer of 1986. PV05 was assigned to flying duties in support of the LHX/ARTI programme.

PV17, 19, and 20 were the first production Apaches assigned to the Fort Eustis, home of the Army Aviation Logistics School, and the base responsible for the training of crew chiefs, maintenance personnel, and maintenance test pilots. The first arrived at the unit on January 30, 1985, joining PV03 which had arrived earlier.

Fort Eustis houses a new 55,000sq ft (5,100m^2) training facility, and normally has 16 Apaches, including ten fully flyable aircraft used as ground maintenance trainers. In this role they allow ground crew to practise working on the type, removing and fitting components and practising other servicing procedures. The remaining six support the Maintenance Test Pilot Program. All pilots initially qualify at Fort Rucker, but those earmarked for flight test duties then attend Fort Eustis for a specialised course lasting more than two months.

Personnel requirement

Each year the US Army requires almost 900 new maintenance personnel to support Apache. Some idea of the magnitude of the training task carried out at Fort Eustis can be gleaned from the fact that McDonnell Douglas Helicopter provided almost a third of a million pages of training material to the base.

The TRADOC training task required the provision of simulators and other

Above: Apache has strong links with the American Indian. It is named after one of the best-known tribes, while production aircraft are flight tested over Indian land near Mesa.

specialised equipment in order to teach new skills to aircrew and ground personnel. In the past the Army has developed its aircraft, then assigned the task of providing training aids to companies which specialised in such work, with the result that during initial training on the type such equipment has not been available, increasing training problems. For the much more complex Apache, however, training aids were wanted ahead of the first production aircraft, in order to support training activities. The intention was that initial service aircraft were to be accepted by crews already trained using ground aids, and to accomplish this the task of providing these equipments was entrusted to the prime manufacturer.

Several types of full-size AH-64

Below: Thirteen patterns of Classroom System Trainer are used to instruct personnel in systems operation and troubleshooting. The central display panel is animated.

replicas have been devised as training aids. The Armament and Fire-Control Trainer incorporates functional wings, pylons, sensor turret and system electronics, while the Flight Control/Pneumatics Trainer takes the form of a skeletal mockup which gives good access to the flight control systems, revealing controls and actuators normally hidden from view by aircraft structure. Its function components are colour-coded, and may be animated to illustrate the operation in slow motion of the control and stabilisation systems.

The most complex training aid at Fort Eustis is the Composite Trainer, essentially a complete aircraft built in a form which facilitates maintenance training. Armament and mission systems may be mockups, but the electrical, hydraulic, pressurised air, fuel and flight-control systems are all present, while an electric motor replaces the T700 turboshaft engine, powering the drive shafting and rotors; the main rotor has cut-down blades in order to save space. The Composite Trainer is used to support many of the classes to show specialists how the aircraft systems fit together. Instructors can deliberately insert around 120 simulated faults so that trainees can practice trouble-shooting and repair.

The task of teaching Army aircrew to fly Apache is handled at Fort Rucker, and PV14, 15, 16 and 18 were the initial aircraft assigned to the base, the first of a fleet of 32 dedicated to training needs. Pilots attend a three-month course, which includes classroom, simulator and flight training. Four computer-controlled part-task trainers – Cockpit Procedures, Mission Equipment and Emergency Procedures Trainers – are installed at the base. These do not totally simulate actual flight, but allow aircrew to practice routine flying, complex flight manoeuvres and even simulated missile firings. Training on the TADS is carried out in the classroom using a Martin Marietta targeting trainer. This combines production hardware with simulation software and allows operation and maintenance training to be conducted at primary and advanced levels.

The Army is always looking for ways to reduce training costs, particularly with systems as expensive as Apache and Hellfire. In order to train aircrew in the technique of low flying by night, Fort Rucker is equipped with ten AH-1 Cobras of novel configuration. Elderly aircraft retired from front-line service, they have been fitted with the PNVS portion of the TADS/PNVS system. Pilots assigned to the AH-64 fly these aircraft during qualification training to gain experience of using the PNVS. As a lower-cost alternative to Hellfire, the Army is considering the possibility of fitting 2.75in rockets with strapdown laser seekers to create training rounds.

Aircraft Development Test Activity is a test unit at Fort Rucker which flies the unit's 33rd Apache. Known as the 'Lead the Fleet' aircraft, this AH-64 will be kept flying throughout the life of the programme at a pace designed to ensure that it will have more flying hours than any other.

Fort Gordon is the TRADOC unit responsible for avionics and ATE (Automatic Test Equipment) training. Skills taught by Signal Corps instructors include on-aircraft maintenance and the use of the Aviation Intermediate Level ATE. No Apache flying operations are carried out from Fort Gordon: its AH-64 'fleet' comprises two Integrated Avionics Maintenance Trainers which consist of AH-64A communications and navigation systems housed in Apache-like airframes. Location of black boxes and the routing of wiring exactly reflect that of the real aircraft, and instructors use control panels to inject simulated faults which the trainees must track down.

Fort Gordon's classrooms are also equipped with around 75 panel trainers. Electronic devices used to simulate and display the aircraft's electrical, armament and hydraulic systems, these are computer controlled and may be used to demonstrate normal system operation and possible fault conditions.

As the training effort built up to ready the Apache for operational deployment, the production programme was also gathering speed. In the autumn of 1984, US Secretary of Defense Caspar Weinberger approved the order of a batch of 160 Apaches, bringing the total number to be obtained to 675, and the programme was duly allocated around $1.2 billion in FY85 for the first year of full-rate production.

In subsequent years, the Apache buy was increased to 807, but an early plan to allow production to peak at a rate of 144 per year in FY85 and be sustained at that level for the next two years was not destined to be adhered to. One culprit may well have been the massive Federal deficit, and the US Army in fact suggested reducing the FY85 purchase to 112 Apaches as a budget-trimming exercise, an idea that was rejected by Congress largely on the grounds that it would have raised unit price by an estimated $1.2 million.

Instead, Congress opted to stay with the original plan which called for 144 aircraft per year but that figure was never actually attained, peak procurement levels being reached in FY85 (138 aircraft) and FY90 (132 aircraft). Reductions

Above: Individual System Trainers are scaled-down versions of the Classroom System Trainer which allow students to try their hand at troubleshooting and other tasks.

Below: Quick-release fasteners give fast access to the avionics bays, a feature which is greatly appreciated by avionics technicians tasked with servicing Apache's electronics.

Below: Fitted out with his IHADSS helmet, a pilot learns basic Apache flying skills on the Cockpit Weapons and Emergency Procedures Trainer.

Left: Aircraft under assembly at the McDonnell Douglas Mesa plant. Getting the production line up to the full production speed of 12 aircraft per month proved a challenging task for the company.

AH-64 Apache production configuration

These drawings show the current production configuration of the AH-64A Apache, including alternative stores, loads-all-Hellfire, mixed Hellfire/Hydra rocket and ferry tanks. Under present plans all production Apaches will be built to this AH-64A standard, although work is already in hand on the proposed follow-on AH-64B. Over 200 AH-64As are also due to be converted to the 'Longbow' configuration.

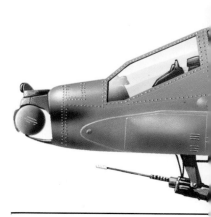

in annual purchases had the effect of extending Apache production and the original plan to cease funding in FY88 also fell by the wayside, with procurement of the original 675 examples for the US Army continuing until FY89, augmented by an additional 132 machines obtained with FY90 appropriations.

US Army AH-64A Purchases by Fiscal Year

Year	Quantity
FY82	11
FY83	48
FY84	112
FY85	138
FY86	116
FY87	101
FY88	77
FY89	72
FY90	132
Total	807

Production of the Hellfire missile built up to match AH-64 deliveries. The original 884-round order in FY82 was followed by a massive 3,971-round order worth $246.3 million in the following year. Production has gathered pace since the early 1980s, with the Army buying 4,651 missiles in FY84, 5,780 in FY85 and 5,750 in FY86 at a cost of around $220 million per year. The Navy was responsible for ordering the missiles needed by the USMC, and its first purchase of 219 in FY84 was followed by orders for 438 and 1,304 in the next two years.

Like most current US defence plans, AH-64 and Hellfire production did not go smoothly. The AH-64 progressed at a somewhat slower pace than originally planned. The FY83 plan called for Hughes and its sub-contractors to deliver 48 Apaches between October 1984 and July 1985. This programme meant that the companies would have to nearly treble their production rate, taking it from three aircraft per month to eight.

Almost inevitably, there were delays. In the summer and autumn of 1984 Teledyne Ryan delivered between two and three fuselages per month, and were the main bottleneck in production. At Culver City, Hughes had problems in accelerating the production rate for some high-strength steel parts, while Martin-Marietta suffered its share of production startup problems with TADS/PNVS. In March 1985, half-way through the 10-month period, production was running at only two aircraft per month. On two occasions, deliveries slipped significantly behind schedule – in the summer and again towards the end of the year.

Bringing the programme up to speed could have been achieved by pumping in extra money, but obviously this would have been costly. Instead, the Army opted to stretch out the FY83 purchase over 14 months rather than 10.

Production acceleration

Getting from eight to 12 per month during 1986 was a demanding target for the production line. 'That last four aircraft [per month] get to be interesting', a company engineer remarked during the author's visit to Mesa in May 1986. Production was then running slightly behind schedule, with aircraft leaving the line at a rate of 10 per month, and due to hit the full rate of 12 per month in July of that year. By the end of 1986 the company hoped to be back on schedule.

The previous year's delivery shortfalls were due to specific problems, the company explained. The summer slippage was due to 'a mishap on the line with the back-up electrical [flight] control system'. In November the delays in production were due to 'an understanding between us and the Army [over] what they wanted on the aircraft... more of a paperwork problem'.

While the company tried to get the line up to full speed, other problems reared their heads. In 1985 the US Army became disturbed by McDonnell Douglas Helicopters' accounting methods, and on May 17 the Army suspended $30 million in monthly overhead payments to the company, a decision announced publicly a week later by Army Secretary John O. Marsh Jr. Undersecretary James R. Ambrose was ordered to investigate 'serious charges of accounting irregularities' by the company.

The mid-1980s saw the Pentagon attempting several reforms of defence procurement. Most publicity was given to instances where military contractors were thought to be overcharging for components and equipment, but an effort was also under way to stamp out abuse of arrangements under which defence manufacturers were allowed to charge the DoD for certain general and administrative costs incurred in weapons production. Investigators had discovered instances of frivolous claims whose total value amounted to hundreds of millions of dollars.

When this area of McDonnell Douglas Helicopters accounts were examined, the Defense Contract Auditing Agency (DCAA) told the US Army that 'the contractor's control of the proper accounting for costs, whether by period, account or pool, cannot be considered adequate. Unallowable costs have been recorded to expense more than once, costs have been recorded in the wrong years, journal entry numbers have been used that were not in the contractor's manual of accounts, and control over the recording of costs by pool, based on organisation codes, is not reliable'.

Inadmissible expenses claimed by the company had included independent research and development, work in preparing bids, plus 'advertising, bad debts, legal expenses, memberships and travel'. One example of duplicated costs was $1,007 million for executives' incentive pay, recorded twice in 1983. Auditors also found large discrepancies in the company's internal records: the DCAA stated that there were no payroll vouchers to support $61 million in employee salaries charged in 1983, or $7.7 million in the previous year.

By June the withheld payments totalled $3,500 million, and the company faced a complete cut-off of funds and suspension of all overhead payments. The DCAA recommended that 'all payments, whether by public voucher or progress payment, will cease by June 17, 1985, if corrective action is not taken, and will not resume until the contractor can demonstrate that all its accounting system deficiencies have been eliminated'.

McDonnell Douglas put on a brave face, pointing out that the practices in question occurred before the 1984 takeover when the company was still part of the Hughes empire. The company had become aware of 'problems and deficiencies' in Hughes Helicopters accounting practices after it purchased the firm, but claimed that even at that stage the problem was being sorted out. Hughes had made 'substantial progress' after starting to correct problems in 1982, said a McDonnell Douglas spokesman, adding that the new owners had accelerated improvements. 'Hughes Helicopters now has good accounting procedures and good internal control standards.'

The Army disagreed. C. Richard Whiston, its chief of legal services, told reporters that audits resulting in the

suspension uncovered problems before and after the sale to McDonnell Douglas. By the autumn the Army concluded the firm had made 'significant progress in revising its accounting procedures', and the suspension was relaxed on August 29.

Two days earlier, on August 27, McDonnell Douglas announced that Hughes Helicopters was being renamed. After 51 years under its existing title and 19 months of existence as a McDonnell Douglas subsidiary, the company would now be known as the McDonnell Douglas Helicopter Company.

The disagreement between the company and the DoD did not involve the actual cost of the aircraft. Apache may be the most expensive attack helicopter ever procured by the US Army, but when examining the current figures it is essential to remember that the original $1.6 million target cost established in 1972 was in then-year dollars, dating from before the vicious inflation triggered off by rising oil prices, and reflected a simpler standard of aircraft. The US Army requirement was revised in 1973 and again in 1976, driven by changing technology and upgrades to the threat which Apache was required to face. To take just one example, the 1972 requirement assumed the use of off-the-shelf TOW missiles and fire-control equipment, but the need for greater stand-off range and fire-and-forget operation resulted in the adoption of the undeveloped Hellfire missile and its associated TADS/PNVS night-vision sensors.

Even given this upgraded specification, Apache's numerical cost increase is similar to that which inflation has imposed on other aircraft over the same time period. The F-5E was originally priced at $1.3 million, but you'd need

Below: Early production Apaches (seen here at Hanchey Army Heliport near Fort Rucker) were used to train the combat pilots who would form the first AH-64 battalion at Fort Hood.

around $8 million in the bank before talking to Northrop today.

In discussing Apache costs, it is essential to adopt the US DoD's own definitions. Unit flyaway cost includes the cost of manufacturing the aircraft and the value of Government Furnished Equipment (GFE) – that is, hardware purchased by the US Government then issued to the aircraft manufacturer, such as the engine and TADS/PNVS – plus project management and support. The US Army target figure is $8.8 million average cost in FY83 dollars for all 675 aircraft currently planned. Actual fly-away cost started high, but has fallen steadily from $10.7 million at the end of 1981 to $8.8 million in late 1984.

Flyaway cost continues to fall. Average figure for aircraft built as part of the FY86 buy was $7.03 million, broken down as follows:

$4.8 million   McDonnell Douglas Helicopters
$0.89 million Martin Marietta
$940,000   Engines (supplied as GFE)
$400,000   Other GFE, for example communications and navaids

Above: The Cockpit, Weapons and Emergency Procedures Trainer is controlled by an instructor, and activated by the electronics racks seen in the background.

Procurement cost adds to the flyaway cost the expenses due to ground support equipment, training and training aids, spares, and support data such as handbooks. Targeted at $11.6 million in FY83 dollars, this has fallen from $11.9 million in FY84 to $8.7 million in FY85.

Program Acquisition unit cost adds a share of the total research and development bill, including the expenses associated with the design, construction and flight test of the YAH-64 and the unsuccessful YAH-63 candidate, plus Hellfire R&D. A US Government report drawn up in April 1985 gave a figure of $13.9 million.

By mid-January 1986 a total of 68 Apaches had been delivered to the US Army, but the company was about to hit a new series of problems, this time technical in nature: on January 15, 1986, inspection of a helicopter which had been set aside for test work revealed a hairline crack in a main rotor blade. The blades had been designed to last for at least 4,500 flight hours, but the one in which the crack had been found had been flown for only about 330 hours.

Examinations were immediately ordered for all Apaches on the inventory, and a further 12 cracked blades were found. Army safety investigators were at a loss to suggest a cause, and on January 27 the Army sent all field units an order to ground the type. The grounding was ordered 'as a precautionary measure', said the Pentagon. 'There have been no accidents related to the main rotor blade'. On the same day, the service decided to suspend Apache deliveries.

Given the testing which the blades had received during development, discovery of the cracks came as an unwelcome surprise for the manufacturer and the user. The main rotor blades had 'demonstrated a capability for continued safe operations after tree strikes or ballistic impact caused by enemy weapons,' the company stated. 'A comprehensive investigation has begun to determine if there is any possible design flaw in this blade, or if there are other factors which contributed to the crack'.

The problem turned out to lie not with the blade design but with one of the tools used in manufacture. This had a slight defect which was causing creases in the trailing edge of the blade, and blade flexure in service resulted in these creases propagating into cracks. The offending tool was modified.

Grounded again

Later that spring a bolt broke in the flight control system of an Apache about to depart on a training mission from Fort Rucker, and the fleet was grounded once again until the cause could be found. Being a critical flight-control component, the bolt had been designed to withstand a direct hit from a 0.5in (12.7mm) machine gun round, so it had been hardened by a heat-treatment process during manufacture, and examination showed that the failed bolt had

Above: The Apache Combat Mission Simulator uses digital simulation to create a realistic view of the external world and armoured targets such as this Soviet T-62.

Below: Simulated TADS imagery on the pilot's head-down CRT display shows a ZSU-23-4 rolling past a church. The "hot spot" on the side of the vehicle shows it is combat-ready.

Above: A close-up of the simulated image shows just how fine the details can be. Despite the horizontal lines, these targets can be identified as ZSU-23-4 Shilkas.

Below: The computer-generated image wouldn't win prizes at an art show, but is detailed enough to give recognisable images of buildings, terrain and other aircraft.

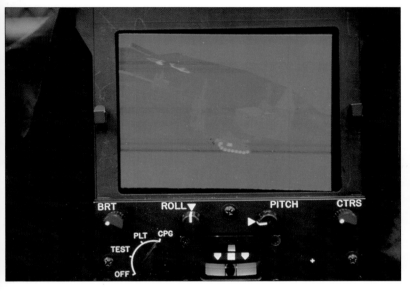

Right: Changes in shading and graduations in contrast all help give clues as to depth and distance, information which strengthens the realism of the scene being depicted.

suffered hydrogen embrittlement. The suspect bolts were removed and replaced by components made from slightly softer material, and subsequent testing of the replaced bolts showed that the specimen which had failed was the only example to show embrittlement.

By the mid-1980s Congress was demanding that major weapon systems be guaranteed by the manufacturer. Apache was one of the first programmes to fall into line, with Hughes guaranteeing that all components which it supplied would be free from material or workmanship defects, meeting or exceeding Army reliability requirements for two years or 240 flight hours, whichever occurred first. If components failed to meet these requirements, Hughes would repair or replace them at no cost to the US Government.

Hellfire problems

Hellfire had its own production problems. By the mid-1980s more than 5,500 rounds were coming off the production line each year, and in FY87 the Army was due to buy 6,576 rounds, the Navy a further 1,364. Then, in a surprise decision, the Army announced that it would not be buying any, shortage of defence funding and minor production problems being given as the reason for the cut. No production money was requested – only $12 million for Hellfire product improvements. No order was placed by the Navy either.

In practice, production lines at Martin Marietta or Rockwell seemed in no danger of being shut down. Missile deliveries had been running between four and five months behind schedule, so a sizeable backlog already existed. Procurement is due to resume in FY88, with the Army ordering 4,800 rounds and the Navy taking 1,824. If the lines show any sign of running out of work before then, the Army has indicated that it would speed up its planned purchase. The Army plans to request $25.3 million in FY88 RDT&E funds for Hellfire product improvements.

Hellfire is expected to remain in service with the US Army and USMC until the end of the century, and under present plans more than 58,000 rounds will be delivered – 47,100 to the Army and 11,139 to the Marines. Total cost of this massive production run has been estimated at $2.5 billion, but there is already good evidence that the programme will come in under budget by around six per cent. When production of Hellfire was authorised in 1982 each missile cost around $63,000, but by 1985 the figure had fallen to $34,000, and the GAO was reporting that FY85 production costs were running 7.5 per cent below target. Assuming that these savings would continue, the GAO estimated that by the time the programme ended a total of $158 million might be shaved off the predicted total bill.

All of the planned operational AH-64 battalions will be formed at Fort Hood in Texas. Some will be AH-1 units converting to the newer aircraft, but under current plans around 10 will be newly-created to deploy the Apache. The first Apaches arrived at the base on March 25 1986. 'It's a great day to be at Fort Hood and in the US Army', said General John Browne, deputy commanding general of III Corps and of Fort Hood, during an acceptance ceremony for the first four aircraft. 'It is the best attack helicopter in the world – none comes close'.

Other VIPs were equally enthusiastic. Apache was 'a truly magnificent piece of equipment that will do mind-boggling things', claimed Army Aviation Center commandant Maj Gen Don Parker, while AH-64 programme manager Maj Gen Charles Drenz announced that 'the book is going to be re-written for Army aviation and the attack helicopter business as we know it'.

The first unit to convert to Apache was the 7th Battalion, 17th Cavalry Brigade (7th/17th) which began its 90-day training schedule in April 1986, the programme being split into three equal-sized 30-day segments. Phase One comprised pilot and maintenance instruction and was basically concerned with imparting a sound knowledge of the

Left: Carefully wrapped in layers of plastic sheeting, this pre-packaged Apache is gently lowered into the hold of a cargo ship. In this way the AH-64 can travel all the way from Texas to northwest Europe.

Apache and its all-important systems to those who would fly and service it. Phase Two saw the Apache crews beginning to operate as teams at company level before moving on to the final stage.

This called for the Battalion to train as a single entity and could perhaps be likened to a graduation exercise, since it did include operations under simulated front-line combat conditions. Since then, apart from minor adjustments arising from experience, the transition process has remained more or less unchanged and quite a few other units have completed conversion at Fort Hood, these typically accumulating about 1,000 hours of flight time during which they will expend thousands of rounds of ammunition and a considerable amount of other ordnance. More importantly, though, after this training period, they will be classified as ready for deployment and active service.

Following hot on the heels of the 7th/17th came the 1st and 2nd Battalions of the 6th Cavalry Regiment/6th Cavalry Brigade (Air Combat). Both units remained initially at Ford Hood although in August 1987 they were ordered to fly their helicopters to Jefferson County Airport on the outskirts of Beaumont, Texas. Once there, the Apaches were transferred to a nearby dockyard and shrouded in plastic sheets before being loaded aboard a pair of US Navy freighters, specifically the USS *Algol* and USS *Capella* which reached the Dutch port of Rotterdam on September 1 and 8 respectively.

The 38 Apaches were only a part of the load, for these two vessels actually ferried a total of 127 helicopters to Europe for Exercise "Reforger '87" (REdeployment of FORces to GERmany) in which they would be required to operate alongside other NATO assets. Reassembly and test flying from Eindhoven presaged a move to the main operating area north of Hanover, West Germany, and once there the Apaches logged some 725 hours in mock combat in what was the type's first large-scale exposure to the European theatre of operations. Night flying figured heavily in the programme as also did adverse weather operations and by all accounts the Apache performed well, with the two units achieving a highly creditable mission-capable rate of over 90 per cent.

On completion of "Reforger '87", the 2nd/6th was destined to remain in Europe, settling into a new home at Illesheim, West Germany while the 1st/6th went back to Fort Hood. Since then, further units have found their way to Europe as the Apache continued to supplant the AH-1 HueyCobra, but it now seems unlikely that US Army plans to base a total of 250+ AH-64As in Europe will be achieved in the light of recent agreements on conventional force levels and the return of some units to the USA after their involvement in Operation "Desert Storm".

For the record other units that were assigned to bases in West Germany in the late 1980s comprised the 4th Battalion/229th Aviation Regiment at Illesheim; the 2nd and 3rd Battalions/1st Aviation Regiment at Ansbach; the 2nd and 3rd Battalions/227th Aviation Regiment at Hanau; and the 5th Battalion/6th Cavalry Regiment at Wiesbaden. Typical battalion strength with an Aviation Regiment consisted of 18 AH-64As plus 13 OH-58C Kiowa scouts and three UH-60A Blackhawks, while battalions assigned to Cavalry Regiments possessed an extra pair of AH-64As.

Modernisation of attack forces has by no means been confined to front-line echelons stationed in the USA and Europe, for the US Army National Guard (ArNG) has also received its fair share of AH-64As fresh from the Mesa production line and more are due. The first ArNG outfit to obtain the Apache was the 1st Battalion/130th Aviation Regiment at Raleigh-Durham Airport, North Carolina. This unit underwent training at Fort Hood in the autumn of 1987 prior to taking delivery of its first six machines on November 9, 1987, and the full complement of 18 had been received by mid-December of that year. Since then, units in South Carolina, Florida and Utah have followed suit.

Brisk production rates were obviously a vital factor in the Army re-equipment programme. After the somewhat shaky start alluded to earlier, once the manufacturing process got into its stride, there was no holding back at Mesa and this in turn permitted unit conversions at Fort Hood to forge ahead rapidly. The pace of that progress is perhaps best exemplified by looking briefly at acceptance rates, starting with delivery of the first production Apache to the US Army on January 26, 1984. It was to take almost 18 months before aircraft number 100 was handed over on June 23, 1986 but subsequent century milestones were passed with fairly constant monotony, with number 200 following on March 2, 1987; number 300 on December 7, 1987; number 400 at the end of October 1988 and number 500 on September 15, 1989.

Self-deployment to Europe

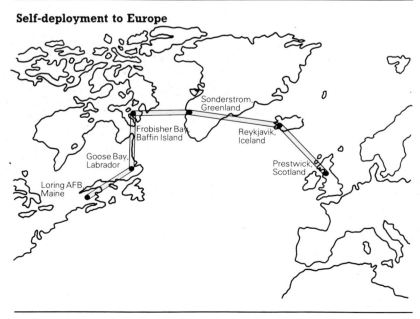

Left: Fitted with external fuel tanks, Apache can cross the North Atlantic in five stages. In an emergency, aircraft could be rushed to Europe by this route, providing swift reinforcement for NATO's front line in West Germany.

Left: An Apache bound for Fort Rucker is loaded into the cargo bay of a C-5A Galaxy. Carried out in June 1985, this exercise confirmed that the Galaxy could carry six Apaches, and proved the preparation and loading procedures.

Air transport configuration

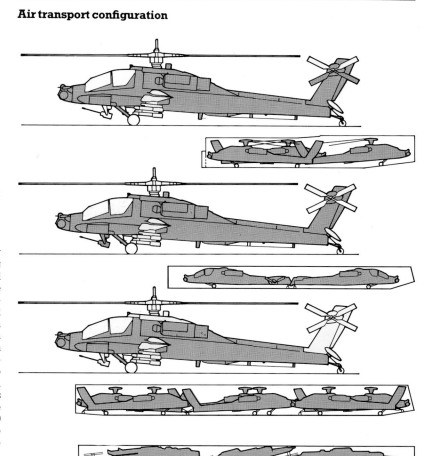

Above: Some items must be removed (pink) or stowed (blue) to allow the shipment of Apache in the C-17 (top), C-141B Starlifter (centre) or C-5A Galaxy (bottom). This work takes one hour per aircraft, or three in the case of carriage by C-141B.

Below: With its main rotor, tail rotor and stabilator removed, and the stub wings folded, an Apache is backed into the cargo hold of a C-5 Galaxy. A seven-man squad can prepare and load six Apaches into a Galaxy in 42 man-hours of effort.

Link simulator

The task of training Apache pilots and gunners to work together as an effective and combat-ready team is made easier by the availability at Fort Hood of the most sophisticated pattern of AH-64A simulator. Requests for Proposals for such equipment were released in January 1982, and in July of that year the task of developing the system was given to the Link division of Singer, which was awarded a $29.7 million contract covering the design and manufacture of a prototype.

Main components of the resulting Combat Mission Simulator are the pilot and co-pilot/gunner trainee modules, realistic replicas of the front and rear cockpits. Both are mounted on a six-degree freedom-of-motion base, and have visual subsystems able to reproduce the external view, plus the imagery which each crew member will see on his video displays.

The system is computer-controlled from an instructor's station located behind each trainee module. Pilot and co-pilot/gunner can train individually or work together as a team in order to 'fly' a simulated combat mission. The unit allows aircrew to practise normal and emergency flight procedures, sensor operation, and weapons delivery. Simulated imagery is available for the FLIR, LLTV and direct view optics, and the pilot's IHADSS HDU and panel-mounted video display unit are fully functional, as are the gunner's IHADSS HDU, and TADS head-up and head-down displays.

This level of complexity is expensive: research and development, plus the construction of a prototype simulator, cost almost $52 million, and estimated cost of the planned production run of simulators is $179.9 million, to which must be added $26 million for related military construction.

Deliveries are due to start in late 1986 with the first batch of three simulators, to be installed at Fort Hood in Texas, and at Wiesbaden and Illesheim in West Germany, and the final three are due to be handed over by the end of 1987. These units should dramatically reduce the cost of AH-64 training: operating cost of the helicopter is at least $2,000 per hour, but simulated mission time will be as little as $300 per hour, so that by the time simulated flying hours pass the 150,000 mark the six units should have come close to paying for their initial cost.

Forward bases

In combat Apache would probably operate from Forward Area Rearming and Refuelling Points (FARRP) located behind the forward edge of battle. The Army hopes that in most cases an AH-64A could be refuelled, rearmed and airborne again within a quarter of an hour – the level of performance which ground crews were able to attain with the earlier Bell AH-1 HueyCobra. A two-man ground crew should be able to reload the aircraft's Hellfire launchers in 5 minutes, or the 2.75in rocket launchers in 10 minutes, while replenishment of the 30mm ammunition should take 10 minutes and the single-point pressure refuelling system is intended to allow the tanks to be refilled in only 4 minutes.

Target mean time between failures is 17 hours. At the flight line all avionics faultfinding will be tackled by the Fiddle built-in test equipment, and line replaceable units will be tested by the AVUM (Air Vehicle Unit Maintenance) trailer, one of which will be issued to each AH-64 company. Faultfinding to printed-circuit card level will be done at battalion level using the RCA-developed Air Vehicle Internal Maintenance (AVIM) trailer.

Apache's sophisticated sensors and displays introduced aircrew – many former AH-1 fliers – to new levels of complexity and to new roles. 'Apache is different than most helicopters,' Warrant Officer Randy F. Dyer, an instructor pilot, explained to a *Washington Post* journalist in the summer of 1986. 'We had to take these aviators and make them system managers.'

Apache crews have to learn two interrelated tasks. Firstly how to use the AH-64, its sensors and weaponry to detect and kill hostile targets; and secondly how to survive in the face of the threats which will in turn be hunting them. In the words of Maj Gen Ellis D. Parker, commander of the Army Aviation Center at Fort Rucker, 'Anything that can be seen can be hit, and anything that can be hit can be killed at unprecedented ranges.' This works both ways, so Apache must spend as much of its combat time as possible in the role of hunter, and the minimum of time as the potential quarry.

OH-58D collaboration

In action Apache will operate in conjunction with the Bell OH-58D Scout helicopter, a rebuilt version of the successful OH-58A Kiowa developed under the US Army's AHIP (Army Helicopter Improvement Program). Most important change to the aircraft is the installation of a McDonnell Douglas Mast-Mounted Sight (MMS) in a spherical housing positioned at the top of the rotor mast.

Manufactured from carbon fibre and located 32in (81cm) above the rotor plane, the 25.5in (65cm) sphere houses a TV camera, thermal imager and laser designator. The FLIR is based on a US Army standard module, and has two

Left: Checkout lists are consulted while others lie stacked on the canopy sill while US Army crews learn the art of operating Apache. The fastened harness suggests that the aircraft is about to be moved.

fields of view – 10° (wide) and 2.8° (narrow). It shares an optical path with the Northrop NdYag laser designator. The TV system operates at a near-infrared wavelength of 1.02 microns, and has 2° and 8° fields of view.

The sphere's stabilisation system has an accuracy of better than 20 milli-radians, and the three sensors can be automatically boresighted in flight in less than 30 seconds. Total weight of the sensor package is less than 150lb (68kg), well below the 280lb (127kg) of the Apache's TADS. A defective sensor can be changed in 15 minutes, the entire pod in 30 minutes.

A flight of five Apaches armed with Hellfire missiles will probably cooperate with three OH-58Ds, and when stalking targets the latter will make maximum use of terrain masking, trying to expose only their sensor balls. Apache would lurk further to the rear, out of range of enemy ground fire.

Typical OH-58D tactics involve positioning the helicopter behind cover, such as a treeline or a fold in the terrain, so that only the sensor ball is exposed to enemy observation. MMS is proving an effective target sensor: early trials showed that target detection range was 22 per cent better than specification using the FLIR, 30 per cent above specification using the TV system.

Once a target has been selected, and its image displayed on a cockpit-mounted CRT, aircrew use a control stick to align the image with the bore-sight of the MMS optical system. The laser is then turned on to designate the target, and information on the victim's position is transmitted back to the waiting Apaches.

The two aircraft must synchronise their operation, ensuring the target remains marked by laser energy throughout the flight time of the Hellfire round. As British Harrier pilots discovered when conducting laser-guided bomb attacks in the closing stages of the Falklands War in 1982, it is easy to get the timing wrong, depriving the smart weapon of its guidance cues. Initially the exchange of data between the OH-58D and AH-64A will be done by secure radio, but an automatic data link is planned.

First launch of a Hellfire missile by an AH-64 working in conjunction with an OH-58D Scout took place at the Yuma Proving Ground in Arizona on August 28, 1984. For the final trial in the Scout development test programme, the crew of the OH-58D illuminated the tank target with the laser and the Apache fired a live-warhead Hellfire from medium range. A direct hit was scored, destroying the target. 'It was a perfect tank kill', announced Lt Col Donald Williamson, assistant supervisor of the Scout programme office. The successful shoot cleared the way for the next phase of Scout operational testing to begin. The Army has asked Rockwell to examine the use of the Aquila RPV as an alternative designator for Hellfire.

Maximum range of Hellfire is around 3.7 miles (6km), a significant increase over other current systems – the Hughes TOW can be used against targets up to

Left: Any open stretch of terrain can be used as an impromptu Apache front-line operating location. Aircraft would probably be kept under camouflage when being worked on, emerging when ready for takeoff.

2.3 miles (3.75km) away, while the Franco-German HOT has a maximum range of around 2.5 miles (4km). The only Western air-launched tactical missile which out-ranges Hellfire is the much heavier Hughes AGM-65 Maverick, which has a range of up to 10 miles (16km), but with a price tag of between $50,000 and $75,000 Maverick is an expensive tank-killer. The Soviet AT-6 Spiral which arms the Hind-E version of the Mi-24 is also a laser-homing weapon, but its range is thought to be around 3 miles (5km), less than that of Hellfire.

Hellfire options

The Apache crew can fire their Hellfire missiles in six different operating modes. Simplest operating method is Lock-On-Before-Launch (LOBL) mode. The seeker of the weapon is locked on to laser energy reflected from the target, then the round is fired. In the basic form of LOBL attack, a single target is assigned to a single missile. Designation can be carried out autonomously by the laser contained in the Apache's TADS system, or remotely using the OH-58D MMS or a ground-based designator operated by combat troops.

If several ground or airborne laser designators are available each can be set to a different code used to designate a separate target, allowing Apache to lock a round on to each target then ripple-launch the missiles at intervals of around one second. If only one designator is available – either Apache's own TADS or a remotely-located laser on another aircraft – the missiles are launched in rapid-fire mode at approximately eight-second intervals.

For maximum firepower against a good concentration of targets the two techniques can be combined, with multiple rounds being ripple-fired and the individual designators switching to fresh targets as soon as the existing ones are destroyed so that a new burst of rounds can be ripple fired. If the Apaches and designators can survive the attention of enemy anti-aircraft systems the process can be repeated until the supply of missiles or targets is exhausted.

Lock-On-After-Launch (LOAL) operating mode has several uses. Its most obvious application is to allow Apache to fire missiles while still masked by terrain or in defilade; in this case a remote designator would probably be used. After launch the missile climbs away along a pre-selected high or low-level trajectory – the former being used to fly over high terrain or obstacles, and the latter in the presence of a low cloud ceiling or as a means of flying over low obstacles – and once a direct line of sight has been obtained to the target the missile locks on and begins to home. Like LOBL mode, LOAL can be used for rapid or ripple fire.

A less obvious application of LOAL is as a method of coping with poor visibility. If the gunner can see the target via TADS but the missile seeker is unable to lock on, the round may be fired in the correct general direction and allowed to lock on once it obtains a strong enough signal.

Apache in combat

Conceived at a time when the Warsaw Pact seemed to pose the most likely threat, the chances of Apache being called upon to employ its weaponry in earnest in the European theatre have diminished greatly as a result of East-West rapprochement. However, there are other areas in the world which are less stable and the Apache has been committed to combat action in two of these areas.

Its first exposure to the hazards of

Below: The OH-58D Mast Mounted Sight contains stabilised FLIR, TV and laser designator systems, allowing the aircraft to shelter behind terrain features while searching for and designating targets for Apache.

Above: Apache will normally operate in conjunction with the Bell OH-58D AHIP scout helicopter, lying some way back from the combat zone and relying on the smaller aircraft to detect and designate targets.

Below: Imagery from the MMS is displayed on a CRT in the OH-58 cockpit. Tests have shown that target detection ranges using the TV and FLIR are well above the figures specified by the US Army.

combat entailed operations on the USA's "doorstep", when American forces were despatched to Panama in Operation "Just Cause" during December 1989. As part of the substantial US Army contingent, a batch of 11 Apaches from the 1st Battalion/82nd Aviation Regiment was transported to Panama aboard C-5 Galaxy freighters of the Military Airlift Command, and these were quickly in action once US forces moved to oust General Manuel Noriega from power and replace him with a democratically elected government. Combat opened with a series of attacks on key military establishments shortly before 0100 hours on

Above: Whilst serving in the war against Iraq in 1991, the Apache was discovered to develop some maintenance problems caused by the fine sand of the region. This resulted amongst other things in a daily flush of the aircraft's engines.

Below: The adage that there is no such thing as an old, bold pilot isn't strictly true in the world of nap-of-the-earth combat. If they are to have any chance of survival, Apache pilots must be prepared to fly at least this close to the ground in tougher tgerrain.

December 20 and the Apaches of Task Force "Wolf" were kept busy between then and the ceasefire which came into effect on January 9, 1990.

During this period, they recorded 247 combat hours, expending a considerable amount of ordnance in direct attacks on a variety of targets. Amongst that ordnance was the Hellfire missile which was also making its combat debut, instances of use including attacks against Noriega's headquarters in which the weapons were fired through selected windows from ranges of up to two miles (4km) with often devastating effect. Hellfire was also given an opportunity to demonstrate its anti-armour potential by accounting for at least two armoured personnel carriers. In addition, the package of night vision aids was also put to good use on armed reconnaissance tasks and the Apache proved of great value in guiding US ground forces around the city as they sought to quell armed resistance.

Operations in Panama

Subsequent Army analysis of the part played by the Apache in "Just Cause" was generally favourable although it did provide some ammunition that was gratefully seized upon by those who were critical of the programme. Instances of operational mishaps appear to have been fairly few but did include the grounding of one machine by hydraulic failure in the lead-up to the assault. This difficulty was compounded when the replacement Apache sustained a loss of TADS imagery after a circuit breaker was tripped as a result of vibration induced by firing of the M230 Chain Gun. Yet another Apache experienced jamming of the gun after an enemy bullet struck the feed mechanism, but all three machines were quickly restored to fully operational status.

Another Apache which suffered 23 hits from Panamanian small arms fire probably took a little longer to repair, especially as it sustained damage to a gearbox as well as the tail rotor drive shaft and the main rotor blades. Fortunately for the crew, the Apache's built-in survival features allowed them to return to base and this ability to sustain battle damage and continue flying must have been most encouraging to the US Army, as would have been the 81 per cent mission-capable rate that was achieved during "Just Cause".

In terms of scale though, the Panamanian adventure was very much a low-key event and was perhaps most valuable in that it allowed the Apache to test the waters of combat. Just over a year later, the AH-64A was to face a much sterner trial of its potential and there can be little doubt that it emerged from Operation "Desert Storm" with a greatly enhanced reputation as a weapon of war.

Deployment of the Apache to Saudi Arabia began within days of Iraq's annexation of Kuwait in August 1990, with elements of the 82nd Aviation Regiment from Fort Bragg, North Carolina, and the 101st Aviation Regiment from Fort Campbell, Kentucky, very quickly being airlifted to Dhahran as part of the US effort to forestall further military adventures by Saddam Hussein. In subsequent months, as the "Desert Shield" build-up continued, still more Apaches were to be moved to the region. Most were withdrawn from US Army assets in Germany in a transfer effort that was unofficially known as "Deforger" and amongst the Apache units that contributed were the 1st Aviation Regiment at Ansbach, the 6th Cavalry Regiment at Wiesbaden and the 227th Aviation Regiment at Hanau.

Early training activity in the desert revealed some problems, especially with the ever-present sand, but modifications to procedures and the availability of hard-stands from which to operate eradicated the worst of these difficulties. As levels of expertise in desert flying rose, so too did operational ready rates improve to around the 90 per cent mark, a level that was maintained for much of the time even after the fighting started.

For the Apache and some of its crews, the transition from peace to war came within hours of the air offensive opening on January 17, when eight AH-64As flew into Western Iraq to attack two key early warning radar sites. In a brief but devastating engagement, no fewer than 27 Hellfire missiles were launched with the aid of night sights and both radar facilities were rendered inoperative, thus clearing a safe corridor for use by coalition fighter-bombers as they headed for their targets deep in Iraq.

Left: The AH-64 was in combat throughout the entire time of Operation Desert Storm. The work of the ground crews was invaluable.

Above: In the background among the AH-64s, somewhere in Saudi Arabia, is the Apache's versatile forward scout: the Bell OH-58D.

the next attack would come from. Indeed, TV reporting of the war included an extraordinary piece of film footage showing Iraqi troops running away from their vehicles to seek cover in the desert when they were fired on by a clutch of Apaches. Within a matter of minutes, the AH-64As wreaked havoc on the column of vehicles with a mixture of gun and missile armament in what was evidently almost a copy-book encounter.

No less extraordinary was the establishment of a forward operating base at Salman airfield, some 50 miles (80km) inside Iraq within hours of the opening of the ground offensive on February 24. Given the code name "Cobra", this airfield was the initial destination for approximately 300 US Army helicopters that included examples of the UH-60A Blackhawk, CH-47D Chinook, OH-58C/D Kiowa, AH-1 Huey-Cobra and, inevitably, the AH-64A Apache. In the remaining hours of the ground war, the base at "Cobra" served as a forward area refuelling and rearming point (FARRP) for many of the attack helicopters and greatly enhanced their productivity by reducing the amount of time required to reach the field of battle.

Apaches were also almost certainly involved in what proved to be one of the war's last major acts, when Iraqi troops fleeing northward from Kuwait to Basra were caught at Mutla Ridge and decimated by US air power. In what became known as the "turkey shoot", thousands of vehicles, including trucks, buses, cars, tanks, armoured personnel carriers and even bulldozers and fire engines, were transformed into little more than scrap metal by a relentless aerial bombardment which turned the three-mile (4.8km) stretch of highway into a charnel house. Within hours, President Bush announced the cessation of hostilities and for the Allied forces and the Apache crews, Operation "Desert Storm" was over . . .

Below: Like a predatory spider lurking near the edge of its web, Apache awaits the approach of its victim. Its web is electronic – radio equipment linking it to ground troops and scout helicopters.

However, it was as a tank-killer that the Apache was designed and it was in this role that it really showed its mettle, statistics released after the war revealing that it was responsible for the destruction of more than 1,000 tanks and armoured fighting vehicles. Working in close co-operation with US Air Force A-10A Thunderbolt IIs as well as US Army and Marine Corps AH-1 Cobra gunships, the AH-64A helped to decimate enemy armour in the vicinity of the border in the early part of the war, before the hunt for fresh targets required them to range farther afield.

Battle for Khafji

Apaches were certainly active in the battle for Khafji, when some 1,500 Iraqi troops supported by about 80 tanks and other armoured vehicles undertook what transpired to be an isolated but determined foray into Saudi Arabia in late January. Fighting was often intense as the Allies sought to recapture the town and air power played a key part in their eventual success, the unfortunate Iraqi troops sustaining a real battering over a period of almost two days as aircraft and helicopters operated virtually unopposed in the skies above.

One of the more remarkable incidents – and one that verified in no uncertain manner the amount of respect that the Iraqi troops had for the Apache and other helicopter gunships – took place on February 17 during a series of skirmishes along the Saudi border area. Then, two AH-64A crews that were busy working over a target were amazed to see some 20 or so Iraqi troops emerge from hiding places clutching white flags as they chose to surrender. Unfazed by this startling development, the two helicopters duly herded the small gaggle of men back towards the American lines where they were taken into captivity.

Other Iraqi soldiers were less fortunate and there can be little doubt that the Apache exacted a heavy toll, especially at night when the ill-equipped Iraqis seemed to have little idea as to where

Performance and Handling

The combined torque of two T700 engines gives Apache the power needed for nap-of-the-earth flight and the hard manoeuvring dictated by combat operations. McDonnell Douglas test pilot Steve Hanvey probably has more Apache flying hours than any other, and was only too willing to describe the flying qualities of the West's hottest helo gunship – an aircraft which he has demonstrated many times for VIPs, the military, and air show audiences. But performance of a modern weapon such as Apache is not just expressable in terms of airframe capability: weapons capability must match it.

Apache is breaking new ground for the US Army, as crews transition from the well-proven AH-1S to the new aircraft. In addition to coping with the new aircraft and its systems, aircrew must learn how to exploit new levels of manoeuvrability and excess power, plus the ability to fly and fight at night and in bad weather.

With the US Army only six weeks into its tight training schedule at Fort Hood when this book was researched in the early summer of 1986, the last thing they wanted was the author interrogating trainees and instructors on Apache's performance and flying characteristics. Luckily, an even more expert Apache pilot was only too happy to extol the aircraft's virtues – Steve Hanvey, head of McDonnell Douglas Helicopter Engineering Flight department.

In theory, the discussion should have been a public relations dream – the company chief test pilot being asked how good his company's product was. Instead the event became more an aviation buff's dream; instead of reciting performance figures and public-relations superlatives, Hanvey spelled out what these meant in practice, the plus factors which Army crews were working to master even as we spoke.

Ground operation

Apache was initially designed to operate from lateral slopes of up to 12°, but a reduction in the length of the weapon pylons allowed this to be increased to 15°. The undercarriage is tall, and the main gear is of narrow track: combined with the soft oleos fitted to absorb the shock of crash landings, these features can make ground taxiing difficult for the neophyte. Pressure on the rudder pedal creates thrust from the tail rotor which, being located high on the tail, tends to make the aircraft lean away from the turn, a sensation which can disconcert first-time riders. Such is the power of the Apache tail rotor that movements of the pedals while the aircraft is stationary with engines running can compress the undercarriage struts.

'The first time you're in a new aircraft, there are certain things you must adapt to,' argues Hanvey. 'You've got a high centre of gravity, but you've also got the very soft ride of the main landing gear... The combination of those two – you can feel that, particularly if you're taxiing too fast. We have never experienced any problems in training. People see that the first time, and then they accommodate it.' The trick is to use the cyclic control to 'fly' the aircraft while still on the ground.

Normal combat weight for the AH-64 will be around 14,700lb (6,660kg). In drawing up its AAH specification, the US Army asked for a 450ft/min (137m/min) vertical rate of climb to be available in typical Middle East conditions – 95°F (35°C) and 4,000ft (1,220m) pressure altitude – when the aircraft was carrying eight Hellfires plus 320 rounds of ammunition. Use of the T700 gave Apache almost three times the specified vertical performance under such conditions, a maximum of 1,450ft/sec (440m/sec). The aircraft can be flown at heavier all-up weights while still meeting the Army's requirements: in the Middle Eastern conditions specified vertical climb rate does not drop to the Army-specified 450ft/sec until 1,200 rounds of cannon are installed rather than the required 320.

Apache can carry more than 5,000lb (2,250kg) of ordnance – more lifting

Above: By opening the throttles and pushing the nose downward, the massive power reserves of Apache's twin T700 engines can be translated into forward acceleration.

Below: During winter manouvres held in early 1986 near Flagstaff, Arizona, an Apache loaded with 2.75in rockets and Hellfire missiles flies out to its assigned combat position.

capability than current weapons can utilise. In combat, the aircraft can be rearmed and refuelled as required to meet mission requirements without being constrained by aircraft lifting performance, enabling commanders to order helo units to 'mix and match' armament and fuel loads as required.

Apache's long range is an important tactical plus. The US Army range requirements call for an 800nm (1,482km) unrefuelled range against a 10kt (18.5km/hr) headwind. This was amply demonstrated on April 4, 1985, when PV14 flew a 1,175 mile (1,890km) sortie in eight hours. The aircraft, fitted with four 230-gallon external tanks, took off from Mesa at 0720, flew directly to Santa Barbara in California, returned to Mesa, flew on to Tucson, Arizona, and returned to Mesa, where it landed with a 30-minute fuel reserve.

Long-distance delivery

A similar exercise was carried out two months later by Army pilots when Fort Rucker received its fourth Apache in June 1985. First leg of the delivery flight was a 535 mile (861km) trip from Mesa to Reese Air Force Base, near Lubbock in Texas. Aircraft and crew remained at Lubbock overnight, then flew 840 miles (1,352km) non-stop to Fort Rucker. Four 230-gallon external tanks were carried for the trip.

In an emergency, aircraft could be fitted with two underwing fuel tanks plus weapons on the other two hardpoints, then flown straight into action at the end of a long delivery flight of 500nm (925km) or more, giving ground troops the fire support needed to deter a would-be aggressor hoping to strike before reinforcements could arrive and prepare for action.

In level flight the aircraft can accelerate at 10kt/sec (18.5km/hr/sec). Apache slightly exceeds the Army cruise speed requirement; maximum continuous engine power normally corresponds to 65 per cent torque, and gives a speed of 146kt (270km/hr). Above this speed drag rises quickly, so that 100 per cent torque takes the aircraft to little more than 150kt (280km/hr).

On most missions the crew have the comforting knowledge that they can survive the loss of one engine, even if the aircraft is in the hover. One of the original two prototypes has demonstrated the ability to hover at an all-up weight of 14,300lb (6,485kg) with 106 per cent

Above: A tilted camera angle adds visual impact to this view of a hovering Apache. Rotor downdraft and engine efflux blur the terrain seen below and behind the aircraft.

torque from a single engine, a figure well within the T700's 115 per cent contingency rating. Height could be maintained at an airspeed of 15kt (0.75m/sec) using 98 per cent torque from a single engine.

An electrical engine cut switch is provided to deal with the effects of any failure of the tail rotor or drive shaft being encountered while the aircraft is hovering at treetop height. If the engine is not shut off immediately after a failure of these components, the main rotor torque will begin to spin the airframe, greatly complicating the severity of the crash, so it is essential that the pilot be able to shut off the engine as quickly as possible. To eliminate the time needed for the pilot to move his hand from the collective control to the throttle quadrant, the emergency engine cut switch takes the form of a collar on the collective control.

Below: Crews of earlier anti-tank helicopters saw their vertical rate of climb eroded by heavy operating weights and hot-and-high conditions. Apache has power and performance to spare, so will have few limitations.

Above: Only friendly forces will get this view of Apache. Enemy forces will rarely if ever see the underside of the aircraft as its crew use its high manoeuvrability in nap-of-the-earth combat tactics.

Below: Speed was not a prime requirement in the AAH specification, but a lightly-loaded AH-64 can still manage more than 160kt (296km/h). Performance degradation in Middle East conditions is negligible.

The flying qualities created by the Digital Automatic Stabilisation Equipment (DASE) impressed *Interavia* editor Mark Lambert when he flew in the front cockpit of the second prototype in 1983. 'The system is optimised for tactical flying, and the degree of control activity needed either to stay accurately in the hover or to move swiftly forwards, sideways or rearwards is astonishingly small... The system responds intelligently to everything the pilot does, with the result that the pilot only needs to "request" a manoeuvre with the stick. Once I grasped the process, I found that the Apache was the easiest helicopter to hover I had ever encountered.' High praise from a man who has flown most types of Western military helicopters.

Even without DASE, Apache is easy to fly. Having flown some hover exercises, Mark Lambert asked Steve Hanvey if he could repeat these with the DASE switched off. To his surprise, he was told

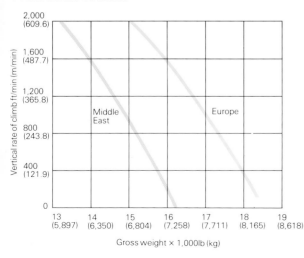

Vertical rate of climb

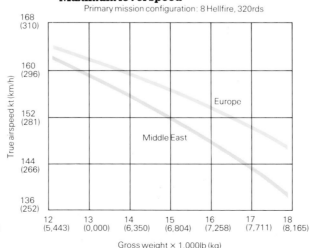

Maximum level speed
Primary mission configuration: 8 Hellfire, 320rds

that it was in fact off, and that he had flown his test manoeuvres without its assistance. 'Because of the location of the trim button, he pressed what he thought was the trim button but it was actually the DASE release,' recalls Hanvey. 'I didn't say anything... so he flew the approach with no stability augmentation. And then he asked if I could take the controls while he made a couple of notes... I asked him if he wanted me to put the DASE back on!'

Time control constant – the time taken to reach two-thirds of the movement demanded via the controls – is fast. Between 120kt and 140kt (222–260km/hr) maximum rate of roll is 100°/sec, close to the maximum acceptable to the US Army, while every inch (2.5cm) of stick displacement gives a roll rate of 26°/sec. 'Roll responsiveness is closer to what I'd describe as like a fighter-type aircraft', says Hanvey. All this is heady stuff for former AH-1 pilots, who are used to rates of only two-thirds this amount.

Pseudo attitude hold modes intended to reduce pilot workload are provided by the DASE. The version used in cruising flight allows the aircraft to fly 'hands off' for extended periods, while in the hover positional data from the Doppler radar and HARS are used to hold the aircraft in position. Any drift due to the aircraft not being completely level may easily be removed by gentle stick movement. Disconnection of the hover mode is automatic as soon as the aircraft reaches a ground speed of 15kt (28km/hr) or a relative airspeed of 50kt (93km/hr).

The sheer power of the Apache's twin engines gives breathtaking performance. Demonstrating the aircraft at the 1983 Paris Air Show, Hanvey used the power of the twin T700s to put the big helicopter through a manoeuvre which seemed almost logic-defying. From the hover, he applied power and climbed to a new hovering position some 300–400ft (90–135m) higher. Flown normally, such a manoeuvre would attract little attention. Flown backwards, it was a show-stopper.

Above: Apache has enough power to carry a full load of weapons plus full internal fuel. As pilots bring their mounts in to land, fuel state should be the least of their worries except after the longest sorties.

Below: Nap-of-the-earth combat flying requires cool nerves and good training, but would be near unthinkable in anything but a highly responsive and utterly reliable aircraft.

The manoeuvre started from a height of around 50ft (15m). 'We were doing about 35 to 40kt backwards, and instead of doing a forward climb all we did was to pull in power to maximum, and climb to stop the rearward flight speed... It's an interesting manoeuvre. Since you're going backwards, to make a climb you push the stick forward... So the tail comes up, and you've got all that energy then going up. Then you pull power to maximum rate of power... So you end up running out of rearward speed. As you come through zero speed, what we do is to just turn out of [the manoeuvre]'.

Flight test to air show

Impressive though it may have been to onlookers, the trick had its origins in a routine flight-test manoeuvre. 'It's just something that developed out of the air-show routine, because you try to stay within a small area. It's a flight-test technique too... there are two ways to stop rearward flight – you can either turn in the direction you're going or you can slowly decelerate. Well, as we got more and more confidence in the aircraft, we'd pitch it down more and more to decelerate. Since that looks kind of funny close to the ground, then we started adding the power in the airshow [routine] to pop it up.'

If this were not impressive enough, he also flew Apache through another odd manoeuvre, spinning the helicopter in the horizontal plane while flying across the airfield – an aeronautical version of the skater's waltz. This manoeuvre was also easier than it looked, he explained, since the axis of the horizontal spin coincided with the pilot's seat. For the front-seater it was a rough ride, however, since he was exposed to the centrifugal force created by the spinning as well as the lateral acceleration caused by the aircraft flight across the field.

'We've seen turn rates of up to 100°/sec while going backwards... translating over the ground at 20 to 40kt', says Hanvey. 'You start in a rearward acceleration, and allow the aircraft to rotate two or three 360s. So of course the

Performance and Handling

Above: Apache's maximum roll rate of 100 degrees per second is more like that of a fighter than a helicopter. Every inch of sideways stick movement adds 26 degrees per second to the roll rate.

Below: With a good degree of down stabilator applied, an Apache crew ease their mount into a desert canyon during a training sortie. Terrain of this type could be met by US rapid-deployment forces.

perspective is – sometimes you're going forward, sometimes you're going sideways, sometimes you're going backwards.'

The exercise had a useful purpose, he explains. 'The demonstration there was to show that [Apache] is not real sensitive to wind from any direction at the higher speeds. Most helicopters at about the 20 to 30kt area, and particularly through the 15kt area, have a lot of control responsiveness problems... The fact that we can do it so fast over such a wide range of speeds shows how well [Apache] hovers in varying wind conditions... I don't concern myself when flying the Apache with winds unless they're over 30mph (48km/h), because I've got plenty of power and never run out of tail rotor [effectiveness]'.

The excess power available from the T700s allows crews to manoeuvre hard when flying nap of the earth, without the risk of losing power and settling into the ground – a potential problem once described by a former AH-1 pilot as 'the biggest thing in the Cobra pilot's mind – thinking about what he's doing with the collective, and what he's asking the ship to do'. This performance is available not only near sea level, but over a wide range of altitudes.

The simplest form of rotor is the rigid type, in which the only freedom given to the blades is the ability to change angle of pitch. This suffers from one major problem – during the portion of the cycle in which a blade moves from the tail of the aircraft towards the nose, it adds its own forward velocity to that of the helicopter, adding to the lift generated. On the other side of the aircraft, blade motion is in the same direction as the airflow past the aircraft, reducing lift. As a result, a helicopter fitted with a rigid rotor tends to roll towards the side with the retreating blade.

Early helicopter engineers struggled with this phenomenon, the solution finally being devised by Juan de la Cierva. He fitted hinges to the roots of the blades of his rotors so that the blades were free to flap above and below the plane of the rotor disc. The increased lift from the advancing blades caused these to rise, reducing lift, while the retreating blades tended to fall, increasing lift. Problems with vibration caused him to add another hinge which would allow the blade a limited amount of freedom in the direction of rotation. The result is known as a fully-articulated rotor.

In an attempt to avoid the complexity of a fully-articulated design, engineers have used modern materials to create the semi-rigid rotor. This has flexible blade roots which act as leaf springs in the horizontal and vertical directions, allowing a degree of freedom in flapping and lag modes, replacing the flapping and drag hinges used in the fully-articulated configuration.

For the UH-1, OH-58 and AH-1, Bell chose a form of semi-rigid rotor often referred to as a 'teeter rotor'. This gives the rotor head a degree of freedom with respect to the vertical drive shaft, so that the aircraft virtually dangles beneath its own rotor. Experience has shown that problems can arise in nap-of-the-earth

flight, when hard manoeuvring can result in the angle between rotor and drive shaft becoming extreme enough to cause the hub to strike the mast. This problem is known as mast bumping, and has resulted in crashes.

Since Apache was intended for low-level flight and hard manoeuvring, the design team opted to stick with an articulated rotor, a configuration which would guarantee a near-fixed relationship between rotor hub and drive shaft and allow the aircraft to be able to accept near-zero or even negative g. Elastomeric dampers provide lead/lag freedom, while offset hinges (that is to say, hinges positioned well outboard of the axis of rotor rotation), cope with flapping motions.

The flight envelope has been cleared from +3.5g to −0.5g, so a pilot flying close to the ground can delay beginning a manoeuvre to match the terrain. A g meter is provided in the cockpit, allowing the pilot to make the best use of low or negative g flight conditions.

'Solid' and 'comfortable' are expressions which come to the lips of Apache pilots when describing their mount's behaviour under g loading, conditions under which some aircraft tend to complain. Hanvey agrees with such sentiments. 'The aircraft feels very comfortable in the two to three g area. Most helicopters tend to resist you in that area – they tend to vibrate badly.'

The ability to pull −0.5g pays dividends under some tactical conditions, minimising aircraft exposure time. 'If someone is looking for you, either with radar or optically, and you come over a crest – that's where you're going to silhouette, and that's where they're going to pick you up. If two aircraft come over at the same time, and both are at full-up weight but one has a faster pitch response which can drive it closer to the terrain before pitching up, it will have less exposure time away from the terrain.'

If flying Apache over a crest during a

Above: This near-plan view of Apache shows the aircraft's fully-articulated main rotor – a design feature which plays a major role in creating high manoeuvrability and good handling in low-level flight.

Below: Apache's flight envelope is cleared from 3.5g to minus 0.5g. The ability to pull negative g allows the pilot to hold the aircraft close to the ground in the dangerous moments after clearing a ridge.

Flight envelope

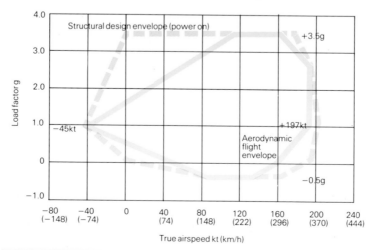

combat mission, '... as soon as I saw that I was clearing the terrain, I would reduce the power... and I'd also pitch over to use the zero g or lower capability. I've waited till I'm closer and I've used the pitch response to hug the ground, then to get up and over. Total Apache exposure time might be five to ten seconds at the most. As anyone who has tried to hoist a man-portable SAM to his shoulder will testify, that's barely long enough to get a shot in. An older and less manoeuvrable helicopter might have anything up to 30 seconds of exposure, enough to let weapons be brought to bear.'

Hanvey has a flight routine which demonstrates this. 'We have a saddle area in the terrain out here [at Mesa], and I'll come up on it. And as I do – it's a sloping terrain – I'll pull Apache up using maybe a couple of gs. As soon as I see the nose start to clear the terrain, I'll push over and reduce the collective, which gets me down to nominally a quarter or even zero g – that's pretty comfortable to the body. Now – another thing you can't do with semi-rigid rotors – I'll roll Apache to stay parallel to the ground, so again I'm staying close to the terrain.'

Performance of Apache cannot be considered simply in terms of how fast, how high and how heavy. A major factor in the aircraft's effectiveness is the capability of the TADS and PNVS: in terms of mission effectiveness Apache is probably more effective at night than it is in the daytime. Apache crews can see and

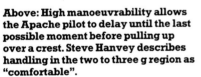

Above: High manoeuvrability allows the Apache pilot to delay until the last possible moment before pulling up over a crest. Steve Hanvey describes handling in the two to three g region as "comfortable".

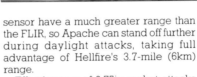

Above right: Once over the ridge of the crest, the aircraft's ability to tolerate negative g allows the pilot to hug the terrain, minimising exposure time to hostile fire from light anti-aircraft weapons and SAMs.

Below: Other attack helicopters head for home when darkness falls, but for Apache it's business as usual. While many threats are blinded by darkness, Apache is not, so can continue to fly and fight.

shoot almost as well at night as in daytime, but most enemy mobile air defence systems are degraded in performance at night. By day, Apache performance may be better but the aircraft faces many optically aimed and manually directed anti-aircraft weapons.

Night flying

With a trained crew, Apache can be flown at night using essentially the same operating procedures and tactics as would be used in the daytime, and crews can locate and identify targets at a much greater range than was possible with earlier aircraft. Crews of less well equipped helicopters who must make do with night-vision goggles have a much poorer night capability, since most goggles rely on starlight to provide sufficient ambient light. Should clouds obscure the sky, the only way of continuing a night attack would be to fire an illumination round, a move which would eliminate any element of surprise.

Using Apache's night-vision systems, crews can carry out Hellfire attacks from extreme stand-off range, relying on darkness to conceal their approach. Unless protected by a suitable warning device such as a short-range radar, an acoustic warner able to detect rotor noise or a laser-warning receiver able to detect the energy from the TADS designator, the enemy AFV formation would be unaware that it had been engaged until the first Hellfire rounds impacted.

The TADS direct-view optics and TV sensor have a much greater range than the FLIR, so Apache can stand off further during daylight attacks, taking full advantage of Hellfire's 3.7-mile (6km) range.

Effectiveness of 2.75in rocket attacks is greater than was the case with the AH-1 Cobra. The Mk 66 rocket motor used on Apache's Hydra 70 rounds gives these a longer range than earlier 2.75in rockets, while the avionics allow them to be aimed with a much higher accuracy and fired from longer range.

Selection of 2.75in rockets automatically activates the appropriate ballistic data for the round to be fired from the information stored in the fire-control computer. The computer also takes inertial body rates from the heading and attitude reference system (HARS), plus information from the air-data system, then calculates weapon-aiming data for the gunner. The 30mm gun has shown that it can destroy light armoured vehicles at ranges not possible with the AH-1's 20mm cannon.

Improving the Breed

Darwin's theory of evolution is just as applicable to warplanes as it is to dinosaurs: evolve or die. As the manufacturer of the Apache, McDonnell Douglas Helicopters certainly got the message at a quite early stage and the design staff at Mesa developed plans for a navalised version as well as improved derivatives for the US Army. More powerful engines, "glass" cockpits and improved Hellfire missiles have all been considered, along with air-to-air armament, enabling the Apache to turn the tables against aircraft that are trying to hunt it down. Having built the world's best anti-tank helicopter, McDonnell Douglas clearly intends to stay ahead of the competition for as long as possible.

Above: Studies of this navalised version of Apache started in 1984. A shallow radome above the rotor mast would house a search radar, and Harpoon missiles could be carried.

With 807 examples on order, the US Army has invested heavily in the Apache and is obviously anxious to explore various ways and means of improving the product so that it might remain an effective and viable asset well into the 21st Century. At the same time, the manufacturer has been no less eager to enhance the product, if only for the simple reason that continuing production means continuing income and anything that keeps the shareholders happy is to be warmly welcomed.

During its production run, the Apache has been able to benefit from new technology whenever this has allowed the creation of better or cheaper components. In the autumn of 1985, for instance, McDonnell Douglas Helicopters was awarded an $806,000 Army Manufacturing Methods & Technology contract to develop and test a new generation of composite materials that would be suitable for use in the AH-64A's secondary structure. These were to be fibre-reinforced thermoplastic materials resistant to high temperature and certain types of solvent. The manufacturer in turn recommended to the US Army that the engine nose gearbox fairings be used to test these new materials, but it is quite conceivable that this research could lead to development of low-cost advanced composites for primary structures in the longer term.

Improved components may also be added to the engine. In March 1985 Garrett Turbine Engine was given an $816,000 contract by the Army Aviation System Command's Applied Technology Laboratory to demonstrate the suitability of laser-hardened heavily loaded gears for aircraft applications. This project is comparing the capabilities of laser-hardened gears against those of gears hardened by conventional carburization techniques. The latter create minor distortion of components, forcing manufacturers to subject the hardened gears to a final grinding process. With laser hardening this final operation could be eliminated, saving time and money. The programme is expected to result in 40 experimental gears being manufactured for the Apache, including the planetary gear for the APU and an accessory drive gear in the main transmission.

One problem with the T700 is that some sections are labour-intensive, consisting of large numbers of parts which must be assembled then inspected. In 1980 General Electric began to study how components of the T700 might in future be manufactured by casting, reducing the number of pieces involved. GE engineers identified areas where casting could be used, and ways in which castings might be produced. 'We took our ideas to the vendors, and said "This is what we want, and we don't know exactly how to do it"', GE project manager Milton Bloom told the magazine *Aviation Week and Space Technology*. 'The vendors then made comments and changes to the drawings, and completed some preliminary castings.'

Each of the new components was designed by a team consisting of a GE engineer plus experts in forging, tooling, and machining, the latter often belonging to the vendors – companies such as Arwood, Howmet Turbine Components, Precision Castparts, and TRW. Goal of the programme was to reduce assembly cost without affecting engine performance or weight, and if the use of casting in any area did not meet these requirements, it was rejected. Progress was not easy; the first cast swirl frame came out 50 per cent overweight.

Compromises often had to be struck in order to make casting practical. Thickness of some components had to be increased, but this could often be offset against the removal of the bolts, flanges

Below: Sidewinder has been tested on Apache, but not adopted for service. Critics of the scheme argue that these weapons might tempt Apache crews to go looking for air combat to the detriment of the main mission.

and doublers required by traditional assembly techniques. Existing assemblies often consisted of several different materials, but a casting had to be manufactured from one material able to withstand worst-case conditions. As a result, the raw material would be more expensive, a factor which had to be traded off against lower manufacturing, assembly, and inspection costs.

A number of areas where casting could usefully be applied were soon identified. These included the front-mounted swirl frame, midframe, integral stage three nozzle and duct, exhaust frame, and bearing support structures, and in some cases the saving was dramatic. The existing midframe consists of around 200 parts whose assembly requires brazing, positioning and electro-chemical machining operations. The new cast design eliminated 80 of these components, with the close tolerances needed for areas such as the anti-ice bleed air vane passages being maintained by ceramic inserts. Casting did not prove practical in every case, but GE eventually decided to adopt the technique for around 60 per cent of the areas originally investigated.

Early in 1985 an experimental engine using cast components was running at GE's Lynn facility. By March it had successfully completed more than 325 hours of endurance testing, and more than two-thirds of the planned 3,300 test cycles due to be carried out before stripdown and post-trials analysis. The Army gave its approval for the use of cast components, and GE hoped to introduce their production within two years.

Maritime proposals

Studies of navalised versions of the Apache effectively began in 1984, although trials had been conducted with a YAH-64 in September 1981 to explore the suitability of the type for deck operation. Two versions were initially proposed – one for the US Marine Corps, which was broadly similar to the US Army Apache, and one for the US Navy. Missions envisaged for the US Marine Corps model included forward air control, gunfire co-ordination for artillery and naval batteries, anti-armour attack, armed reconnaissance, escort of transport helicopters and surface vessels and even anti-air operations.

Features of the proposed design included the use of naval-standard T700-401 engines, the addition of an automatic blade-folding system and a Doppler-inertial navigation system, plus various anti-corrosion improvements intended to help the airframe withstand the

US Marine Corps amphibious support mission

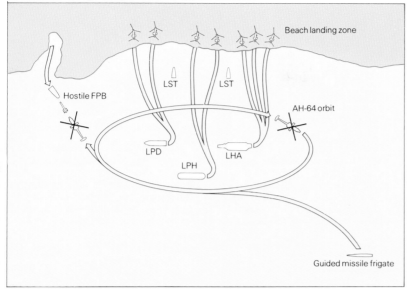

US Navy anti-air warfare mission

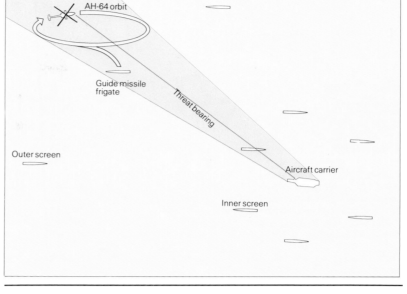

Top: Two possible missions for a navalised Apache: amphibious support would involve helicopters orbiting over the assault ships as the air and shipborne landing forces hit the beach; at the left, one AH-64 engages a hostile vessel. The US Navy might use frigate-launched AH-64 patrols on the bearing of a threat detected by a distant carrier.

rigours of naval service. A blade-folding system was developed by McDonnell Douglas Helicopters and a full-size model built, but plans to test-fly the new components so as to verify that these would pose no dynamic problems were postponed.

Armament options would include those already employed by US Army Apaches, but the Sea Apache proposal anticipated the use of additional ordnance that included AIM-9 Sidewinder and FIM-92 Stinger infra-red homing air-to-air missiles, as well as the Hughes TOW missile and 5in (127mm) unguided rockets for air-to-surface attack tasks. Defence suppression was another mission that was considered, using the Sidearm anti-radiation version of the Sidewinder to take out hostile radars.

Below: The first launch of an AIM-9M Sidewinder by an AH-64. This took place at White Sands, New Mexico in November 1987.

In the case of the US Marine Corps, a typical mission in support of amphibious assault operations envisaged the carriage of mixed armament that included four Hellfire missiles for use against point and armoured targets, 38 2.75in (70mm) rockets and 1,200 rounds of 30mm ammunition for landing-zone fire suppression plus two AIM-9s. In this configuration, estimated figures for mission radius and endurance were 142 miles (228km) and 2.8 hours respectively.

At least two different proposals for a US Navy version are known to have existed. The first (from 1984) embodied a search radar housing above the rotor mast – this would have retained the TADS/PNVS package and was therefore compatible with the full array of US Army-type armament, as well as AIM-9 Sidewinders for self-defence and either Harpoon or Penguin missiles for anti-ship strike duties. One anti-ship attack mission profile anticipated the Sea Apache being able to fly out up to 150 miles (240km) from the parent vessel to search for, identify and, if appropriate, engage enemy surface vessels with either Harpoon or Penguin, but extended range could be achieved by deleting armament in favour of auxiliary fuel tanks.

A later (1987) and somewhat more fully worked out Sea Apache concept was radically different in that it featured a retractable undercarriage, with the main units mounted on outriggers so as to enhance stability when flying from landing platforms. An inflight refuelling probe was also added on the starboard fuselage side, but perhaps the most significant difference concerned the nose section which was drastically revamped. Gone was the TADS/PNVS suite, its place taken by a more conventional radome covering a Hughes AN/APG-65 multi-mode search radar. Gone too was the Chain Gun, armament options being more limited, artist's impressions released at the time showing just Sidewinder and Harpoon missiles, giving a very clear indication that this version of the Sea Apache was intended primarily for anti-ship strike missions.

Proponents of the Sea Apache were perhaps overly optimistic when they spoke of up to 100 examples being required for service with four US Navy squadrons, for the service already had a fairly potent rotary-wing anti-ship attack capability in the shape of the SH-60B Seahawk. So, while the Sea Apache might have figured fairly highly in any list of equipment that would be "nice to have", the existence of the Seahawk and increasingly stringent financial considerations combined to render it a non-starter.

Improving the breed

All Army aircraft have at a tendency to grow in weight throughout their careers as more hardware is added and the Apache is unlikely to be any different, profiting from work in areas such as improved fire-control systems and laser-warning receivers. The aircraft's impressive vertical climb performance provides a useful growth capability, and will inevitably be reduced by the effect of increased weight as the Apache matures.

The Apache does an excellent job in the role for which it was designed, but the US Army's operational requirements and doctrine are not standing still. Like the threat capability of potential opponents, they are constantly being improved with time and maintaining the effectiveness of the contemporary Apache is certain to involve changes and upgrades in years to come.

One agency that is intimately associated with efforts to enhance the Apache is the Army Program Office and this began to conceive ideas for short-term improvements at an early stage in the AH-64A's operational career, although studies soon confirmed that piecemeal implementation of these ideas would be costly and that it would be more sensible to add upgrades as part of an integrated package known as Block Improvements. This concept initially met with a favourable response from the Army and a number of possible Block 1 improvements were considered under an Army-funded project which envisaged an upgraded version beginning to roll from the assembly lines at Mesa with effect from the 516th Apache. As it turned out, the idea lapsed and current production AH-64As differ in only minor detail from their predecessors.

A more promising concept originated from McDonnell Douglas Helicopters in the mid-1980s, when the manufacturer headed an industry team that examined technology for a possible second-generation member of the Apache family. This upgraded version was originally referred to as the Advanced Apache but later evolved into Apache Plus and was for a time referred to unofficially as the "AH-64B".

Intended to introduce such features as automatic fire-control, Stinger air-to-air missiles, a redesigned cockpit, an aft-looking TV with a fin-mounted video camera and a gun with an extended barrel plus digital turret control, this version would have possessed about 75 per cent commonality with the AH-64A, but it failed to progress beyond the drawing board. Equally unsuccessful were attempts to launch a Multi-Stage Improvement Programme (MSIP) which would have entailed upgrading of exist-

Above: The front crew station of an AH-64 has been fitted out with sidearm fly-by wire controls to represent the single-pilot cockpit of the proposed LHX helicopter. Testing began on October 16, 1985.

Above: The main European rival to the Apache is the Eurocopter Tiger. This Franco-German joint venture is expected in service in the late 1990s.

Below: The most visually distinctive thing about the AH-64 Longbow Apache is its large fire-control radar which sits atop the main rotor.

ing weapons, sensor and fire control systems as well as changes to the digital databus and crew displays. Some funds were appropriated for MSIP development in the FY88 budget and development work was launched in mid-1988, but this was soon terminated.

Longbow Apache

However, some of the items detailed above are almost certain to be incorporated in the so-called Longbow Apache which completed a period of proof-of-principle testing in the spring of 1990. A pair of AH-64As were fitted with Longbow radar in order to take part in Early User Test and Experimentation (EUT&E) trials in April 1990, during which they accomplished seven live firings of RF Hellfire missiles, six of which scored hits on moving and stationary tank targets.

Authorisation to proceed with full-scale development (FSD) duly followed, but work on this project has been consciously delayed by some 20 months in order to more closely coincide with development of a new version of Hellfire. Once FSD does get under way, it is planned to last some 50 months, clearing the way for operational deployment during the course of Fiscal Year 1996.

Readily identifiable by virtue of the mast-mounted Martin Marietta/Westinghouse Longbow millimetre-wave radar, this version of Apache will possess enhanced all-weather operating capability. Locating the radar on the mast enables the Longbow Apache to track air targets through a full 360deg field of view and ground targets through 270deg, while compatibility with a radar-homing fire-and-forget version of Hellfire should greatly enhance "kill" probabilities. Known as the Hellfire Optimized Missile System (HOMS), the new model has a revised warhead which is expected to be able to cope with advanced types of armour-plating, while it also embodies a seeker-head that is "hardened" against electro-optical countermeasures.

Emerging victorious from an Army competition against Rockwell International, Martin Marietta was awarded a $10 million contract covering a 27-month FSD effort that will include no fewer than 65 HOMS launches. Starting in late 1991, those firing trials will confirm whether HOMS will go ahead and it is already clear that the potential rewards are high, since the US Army has identified a requirement for close to 11,000.

If it goes ahead, Longbow Apache will be identified by the US Army as the AH-64C and current planning calls for the service to obtain some 227 examples by means of an upgrade of existing AH-64As rather than through new procurement. Other improvements to be embodied evidently include re-engining with 1,875shp (1,385kW) T700-GE-701C turboshafts and provision for Stinger AAMs on wingtip pylons. Stinger is also to be added to another version of the Apache, specifically the AH-64B, which is a less ambitious reworking that will also benefit from provision of an automatic target hand-off system and the Navstar global positioning system receiver.

In the case of the AH-64B, the Army presently intends to convert 580 AH-64As to this configuration. However, the figures mentioned here for the two upgrading efforts will undoubtedly have to be adjusted to take account of attrition and one is inclined to think that it will be the AH-64B version that will bear the brunt of any curtailment that is found necessary.

Below: Inside the Longbow Apache, the cockpit instrumentation will be more markedly different. This is a simulator of the co-pilot's station.

Evasion of fighter attack

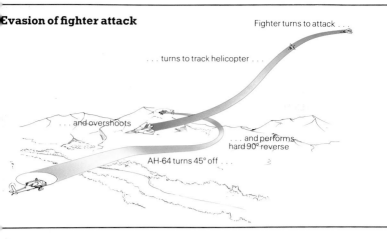

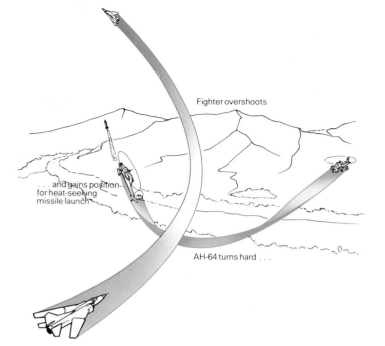

Above: Helicopters such as Apache are no easy prey for a jet fighter. This defensive manoeuvre reduces tracking time and changes the angle off, spoiling the fighter's attack and causing it to overshoot.

Above right: Once the fighter has been persuaded to overshoot, the helicopter can use its turning ability to get into position for a missile shot. The AIM-9s tested on Apache would then prove invaluable.

Right: The smart fighter pilot will attempt to counter this tactic by passing close inboard of the turning helicopter, then manoeuvring hard and dropping to low level in order to deny its opponent a firing chance.

Below right: If the fighter attempts a second firing pass, the helicopter can counter this by steering towards the focus of the fighter's turn, forcing it to attempt an ever-tightening and impractical turn.

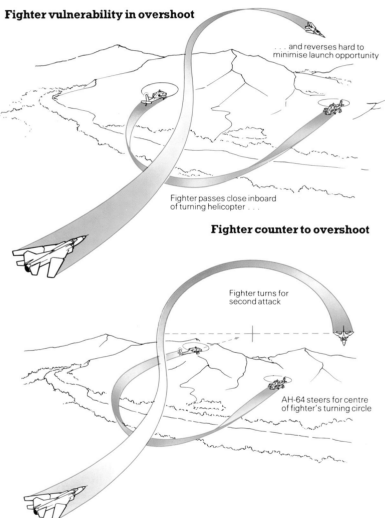

Apaches for export

Returning to the Longbow Apache, there is a strong possibility that this derivative will also be produced in new-build form, for a customer on the other side of the Atlantic Ocean. The matter of finding a new anti-armour attack helicopter to replace the existing Westland Lynx/TOW combination in service with the British Army Air Corps (AAC) is one that has exercised officials of that service and the Ministry of Defence (MoD) for some considerable time. Various options have been considered, but the AAC has never been shy about pressing its desire for a version of the AH-64 and this eventually hardened in favour of Longbow Apache.

In the summer of 1991, the AAC's preference seemed to be very much in the ascendancy following its endorsement by the MoD, but the eventual outcome is far from cut and dried, since the Ministry's Procurement Executive has decreed that Apache must first negotiate a competitive evaluation against the Eurocopter Tiger. Not surprisingly, the AAC is anxious to avoid further delay and it is probably also concerned that politically driven considerations may yet conspire to prevent it from obtaining what it feels is the most suitable type. However, assuming that Apache is selected, procurement is likely to total around 125, with the first examples expected to enter service in the 1996-97 timeframe, coincidentally the target date for deployment of Tiger with elements of the French and German armies.

Whichever type is eventually selected, it seems probable that assembly will be undertaken in the United Kingdom by Westland Helicopters which has hedged its bets by signing memorandums of understanding with both competitors. The odds do at present appear to favour the American contender.

While the outcome of the British order may still be very much in doubt, the Apache scored its first success on the export market at the beginning of 1990 when it was revealed that the Israeli Defence Force/Air Force (IDF/AF) was to receive 18 examples of the AH-64A under US Foreign Military Sales provisions. Following crew training in the USA during the summer of 1990, the IDF/AF commander-in-chief, General Avihu Ben-Nun, accepted the first Apache at Ramon air base on September 13, 1990.

Since then, a number of other Middle Eastern states have been authorised to receive Apache. The first of these customers was Egypt, which is to receive 24 – as with Israel, crew training began in the USA before the end of 1990. The closing months of that same year also saw two more customers emerge in the shape of Bahrain and the United Arab Emirates, which are expected to obtain eight and 20 helicopters respectively as part of a package of military hardware that also includes main battle tanks and other armoured fighting vehicles, as well as Hellfire missiles for the AH-64As.

Saudi Arabia has also expressed considerable interest in obtaining Apaches and was at one time expected to receive 48 as part of a long-term package of military equipment that was hastily conceived at the height of the Gulf crisis. Objections by pro-Israeli factions in the USA forced a reappraisal and a mid-term plan was conceived which would have included a dozen Apaches to be supplied in late 1991. Eventually, even this modest amount was shelved in the wake of the successful prosecution of Operation "Desert Storm".

At the time of writing, no other foreign customers have yet elected to purchase the Apache, but at least two nations are showing definite signs of interest. One is Japan, whose Ground Self-Defence Force (JGSDF) may well opt to buy a handful of AH-64As in 1994-95 for evaluation and operational test duties. While the JGSDF would almost certainly wish to obtain Apache in quantity as a replacement for the AH-1S, the political and financial climate is not exactly conducive to a major order.

There is also room for doubt about the future of a Dutch proposal to convert an Army armoured brigade into an airmobile brigade that would include attack and transport helicopters, even though this has received parliamentary approval. Present plans anticipate a leasing arrangement whereby about 40 attack helicopters made surplus by the Conventional Forces Europe (CFE) agreement would be operated as a lead-in to procurement of new machines. Candidates under consideration for leasing consist of the BO 105P/PAH-1, the AH-1F HueyCobra and the AH-64A Apache, while competitors for new procurement are the Apache and the Eurocopter Tiger. However, Labour elements of the centre-left coalition government are known to oppose these plans.

Future Air Attack Vehicle

Looking even further ahead, the US Army is hoping to field a Future Attack Air Vehicle (FAAV) – which may or may not be a helicopter – during the latter part of the next decade and will have to make a decision in 1997 as to whether or not to proceed with what is clearly an ambitious programme. In broad terms, this concept envisages a fast and "long-legged" machine that possesses air-to-ground and air-to-air capability and at the moment there appear to be two options.

One is to go for an entirely new type. The other is to purchase an advanced and redesigned version of the Apache, with the ability to undertake light attack missions as well as the existing heavy attack task. So, the end of production of the AH-64A for the US Army in 1993 does not necessarily signify the end of Apache procurement for that service, although any FAAV derivative may bear only a passing resemblance to the Apache as we now know it.

Left: Atop the rotor, the Longbow Apache's Airborne Adverse Weather Weapons System will give the future AH-64s the fighting edge.

Specifications

MCDONNELL DOUGLAS HELICOPTER AH-64A APACHE

DIMENSIONS

Length overall, rotors turning	58ft 3in (17.76m)
Main rotor diameter	48ft 0in (14.63m)
Tail rotor diameter	9ft 2in (2.79m)
Wingspan	17ft 2in (5.23m)
Height over tail fin	11ft 6in (3.52m)
Height over main rotor	12ft 7in (3.84m)

WEIGHTS

Empty	10,760lb (4,881kg)
Primary mission gross	14,445lb (6,552kg)
Max take-off	21,000lb (9,525kg)

PERFORMANCE

Never exceed speed	197kt (365km/h)
Max cruising speed	160kt (296km/h)
Max vertical rate of climb	2,500ft/min (762m/min)
Service ceiling	21,000ft (6,400m)
Low altitude g limits	+3.5 to −0.5
Max range (internal fuel)	260nm (482km)
Ferry range with external fuel	918nm (1,701km)

GENERAL ELECTRIC T700-GE-701 TURBOSHAFT

Length	46.5in (1.18m)
Width	25in (63.5cm)
Height	23in (58.4cm)
Weight	437lb (198kg)
Compressor	5 axial stages plus 1 centrifugal
Combustor	annular
Turbine (gas generator)	two-stage
Turbine (power)	two-stage
Output shaft speed	17,000-21,000rpm
Sea-level output power:	
Maximum continuous power	1,510shp (1,126kW)
Intermediate power	1,696shp (1,265kW)
Emergency rating	1,723shp (1,285kW) for 150 secs

ROCKWELL INTERNATIONAL AGM-114A HELLFIRE

Length	64in (162cm)
Diameter	7in (18cm)
Launch weight	95lb (43kg)
Propulsion	Thiokol TX-657 solid-propellant rocket motor
Guidance	Semi-active laser seeker
Warhead	17lb (7.7kg) shaped charge
Range	greater than 3.7 miles (6km)

Performance

EUROPE: AMBIENT CONDITIONS 2,000ft (600m), 70°F (21°C)

Mission	Missiles	Ammunition	Vertical climb	Level speed	Mission duration
Anti-armour defence	16HF	1,200rds	990ft/min	148kt	2.5hr
Covering Force (Air Cav)	8HF/38rkt	1,200rds	860ft/min	150kt	2.5hr
Airmobile escort	76rkt	1,200rds	780ft/min	153kt	2.5hr

MIDDLE EAST: AMBIENT CONDITIONS 4,000ft (1,200m), 95°F (35°C)

Mission	Missiles	Ammunition	Vertical climb	Level speed	Mission duration
Anti-armour defence	8HF	320rds	1,450ft/min	154kt	1.83hr
Anti-armour defence	8HF	1,200rds	450ft/min	151kt	2.87hr
Anti-armour defence	16HF	320rds	450ft/min	147kt	1.9hr
Covering Force (Air Cav)	8HF	1,200rds	960ft/min	153kt	1.83hr
Airmobile escort	38rkt	1,200rds			

Apache production and training

Fort Eustis, Virginia
US Army Aviation Logistics Center: maintenance training

Fort Gordon, Georgia
US Army Signal School: avionics and test equipment training

Fort Hood, Texas
US Army Forces Command Apache Training Brigade: battalion level training

Loring AFB, Maine
Staging post for deployment to Europe

Mesa, Arizona
McDonnell Douglas Helicopter Company: AH-64 assembly and flight test centre

Fort Rucker, Alabama
US Army Aviation Center: pilot training

Picture credits

Endpaper: McDonnell Douglas **Title page:** McDonnell Douglas **Page 2/3:** (all) McDonnell Douglas **4:** (top and bottom) US Army; (centre) McDonnell Douglas **5:** (both) Bell Helicopter **6:** (top) Bell Helicopter; (bottom) McDonnell Douglas **7:** (top) McDonnell Douglas; (bottom) General Electric **8:** (top left) McDonnell Douglas; (remainder) US Army **9:** (both) McDonnell Douglas **10/11:** (top left) US Army; (remainder) McDonnell Douglas **12:** (top) Sikorsky; (bottom) McDonnell Douglas **13:** (top) McDonnell Douglas/TRH; (bottom) McDonnell Douglas **14:** (both) McDonnell Douglas **15:** (top) US Army; ((remainder) McDonnell Douglas **16/17:** (all) McDonnell Douglas **18/19:** (all) McDonnell Douglas **20/21:** (all) McDonnell Douglas **22/23:** (top) US Army; (bottom left and right) McDonnell Douglas (upper centre) US Army; (bottom centre) Fiat Aviazone **24:** (both) McDonnell Douglas **25:** (top) McDonnell Douglas; (bottom) General Electric **26/27:** (bottom centre) McDonnell Douglas (remainder) Martin Marietta **28:** McDonnell Douglas/TRH **29:** Honeywell **30:** (top left and right) McDonnell Douglas; (bottom) DoD/MARS **32/33:** (centre left) DoD/MARS; (remainder)McDonnell Douglas **34/35:** (all) McDonnell Douglas **36/37:** (top centre) McDonnell Douglas/TRH; (remainder) McDonnell Douglas **38:** (top) US Army; (bottom left) McDonnell Douglas **39:** (all) McDonnell Douglas **41:** McDonnell Douglas **42:** (top) US Army; (bottom) McDonnell Douglas **43:** (all) McDonnell Douglas **44:** McDonnell Douglas **46:** (both) McDonnell Douglas **47:** (all) Singer/Link **48/49:** (all) McDonnell Douglas **50:** (both) McDonnell Douglas **51:** (top) US Army, (remainder) Bell Helicopter **52/53:** (bottom left) DoD/MARS; (remainder) US Army **54/55:** (all) McDonnell Douglas **56/57:** (all) McDonnell Douglas **58/59:** (all) McDonnell Douglas **60/61:** (all) McDonnell Douglas **62:** (top right) Eurocopter; (remainder) McDonnell Douglas